T0231465

The Newman Lectures on Mathematics

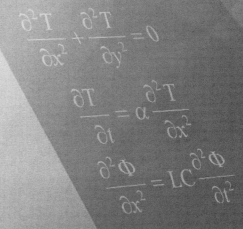

$$\frac{\partial^2 T}{\partial x^2} + \frac{\partial^2 T}{\partial y^2} = 0$$

$$\frac{\partial T}{\partial t} = \alpha \frac{\partial^2 T}{\partial x^2}$$

$$\frac{\partial^2 \Phi}{\partial x^2} = LC \frac{\partial^2 \Phi}{\partial t^2}$$

The Newman Lectures on Mathematics

John Newman
Vincent Battaglia

PAN STANFORD PUBLISHING

Published by

Pan Stanford Publishing Pte. Ltd.
Penthouse Level, Suntec Tower 3
8 Temasek Boulevard
Singapore 038988

Email: editorial@panstanford.com
Web: www.panstanford.com

British Library Cataloguing-in-Publication Data
A catalogue record for this book is available from the British Library.

ISBN 978-981-4774-25-3 (Hardcover)
ISBN 978-1-315-10885-8 (eBook)

Contents

Introduction and Philosophical Remarks

This book covers vector calculus and ordinary and partial differential equations, both with Laplace transforms. However, the treatment of ordinary differential equations is somewhat abstract because many ordinary differential equations of interest arise in the course of solving partial differential equations. Occasionally, a problem is reduced to an integral equation or the utility of numerical or perturbation methods is indicated. Singular perturbations have been detailed further in Volume 3, *The Newman Lectures on Transport Phenomena*, of this book series.

The present book emphasizes that readers should be able to analyze the problems they encounter in chemical engineering courses. It exposes them to methods of mathematical thinking and presents selected examples that illustrate the techniques that are useful in a typical evening by the fire with a pen and a pad of paper. The book does not focus on the rigorous proof of theorems such as existence and uniqueness of solutions. However, it does give importance to formal manipulations because trivial errors consume a lot of time, which can otherwise be devoted to useful activities. It suggests the readers to follow these three important steps for problem solving:

1. Formulate the problem in mathematical terms. This may be done with varying degrees of completeness or detail, but one should always make sure that the important features of the physical situation are adequately described. For problems of common types, this part may be easy, but some problems are new and require careful consideration.
2. Work through to obtain a solution using the mathematical tools at your disposal. You may need to introduce additional approximations or assumptions in order to get an answer.
3. Contemplate the physical meaning of the results. You should be able to explain qualitatively why the results behave as they do. Be on the lookout for physical absurdities or impossibilities. These may result from an incorrect formulation of the problem

with neglect of important factors or from approximations or even errors introduced during the solution. It should be possible to check, at least *a posteriori*, the validity of an approximation. On the other hand, one may want to live with the consequences of an approximation, recognizing the limitations of the solution in certain regions.

Chapter 1

Differentiation of Integrals

In the solution to differential equations, either in abstract terms or in specific instances, one frequently arrives at an integral of the general form

$$I(x) = \int_{L_1(x)}^{L_2(x)} F(x,\xi)d\xi . \qquad (1.1)$$

This is so because the differential equation is considered solved if the expression for the unknown can be reduced to such an integral, even if the integral cannot be evaluated in the closed form.

In order to verify such a solution or to derive its properties, it may be necessary to differentiate it. Hence, you should verify that its derivative is

$$\frac{dI}{dx} = \frac{dL_2}{dx}F[x,L_2(x)] - \frac{dL_1}{dx}F[x,L_1(x)] + \int_{L_1(x)}^{L_2(x)} \frac{\partial F}{\partial x}d\xi . \qquad (1.2)$$

This is called the Leibniz rule.

Problems

1.1 Verify Eq. 1.2.

1.2 Differentiate $\int_{y/2\sqrt{Dt}}^{\infty} e^{-\xi^2} d\xi$ with respect to t.

The Newman Lectures on Mathematics
John Newman and Vincent Battaglia
Copyright © 2018 Pan Stanford Publishing Pte. Ltd.
ISBN 978-981-4774-25-3 (Hardcover), 978-1-315-10885-8 (eBook)
www.panstanford.com

1.3 Differentiate the integral of Problem 1.2 with respect to y.

1.4 Can the integral in Eq. 1.2 be evaluated directly, since it is an integral of a derivative?

Chapter 2

Linear, First-Order Differential Equations

The general form of a linear, first-order differential equation is

$$\frac{dy}{dx} + a(x)y = f(x). \tag{2.1}$$

This can be solved by means of an integrating factor F,

$$F = \exp\left\{ \int_{L_1}^{x} a\, dx \right\}, \tag{2.2}$$

where the lower limit L_1 can be selected to make F as simple as possible. For example, if $a = x$, the selection $L_1 = 0$ gives $F = e^{x^2/2}$. If $a = 1/x$, the selection $L_1 = 1$ gives $F = x$.

Multiplication of Eq. 2.1 by F yields

$$F\frac{dy}{dx} + Fay = \frac{dFy}{dx} = Ff . \tag{2.3}$$

The form 2 of the integrating factor F is easy to remember since it converts the left side of Eq. 2.3 to a perfect differential. To verify this, one needs to make use of Eq. 1.2 for the differentiation of an integral.

Integration of Eq. 2.3 gives

$$Fy = \int_{L_2}^{x} Ff dx + A , \tag{2.4}$$

where A is a constant of integration and where again L_2 can be selected so as to yield the simplest expression. Thus, the general solution for y is

The Newman Lectures on Mathematics
John Newman and Vincent Battaglia
Copyright © 2018 Pan Stanford Publishing Pte. Ltd.
ISBN 978-981-4774-25-3 (Hardcover), 978-1-315-10885-8 (eBook)
www.panstanford.com

$$y = e^{-\int_{L_1}^{x} a\,dx} \int_{L_2}^{x} f e^{\int_{L_1}^{x} a\,dx}\,dx + Ae^{-\int_{L_1}^{x} a\,dx}. \tag{2.5}$$

This general solution contains one arbitrary constant A, as is appropriate for a first-order equation. Here one recognizes the first term in Eq. 2.5 as a particular solution to Eq. 2.1 and the second term as the general solution to the corresponding homogeneous equation, as discussed in the next section on linear systems.

As an example, the general solution to the equation

$$\frac{dp}{dx} + 2xp = 1 \tag{2.6}$$

is

$$p = e^{-x^2} \int_0^x e^{x^2}\,dx + Ae^{-x^2}. \tag{2.7}$$

One should notice that here we have not been careful to distinguish the dummy variable of integration x from the independent variable x. If both appeared in the integrand, one would have to distinguish the two. For example, Eq. 2.7 should, more properly, be written as

$$p = e^{-x^2} \int_0^x e^{\xi^2}\,d\xi + Ae^{-x^2} = \int_0^x e^{\xi^2 - x^2}\,d\xi + Ae^{-x^2}. \tag{2.8}$$

Problems

2.1 Show that Eq. 2.5 is a solution to Eq. 2.1.

2.2 Write the general solution to Eq. 2.6 with $L_1 = 1$ and $L_2 = 2$ and show that the result is equivalent to Eq. 2.7.

Chapter 3

Linear Systems

A linear problem is to determine y from the equation

$$\mathcal{L}\{y\} = F, \qquad (3.1)$$

where F is given and $\mathcal{L}$ is a linear operator. A linear operator has the following properties:

1. $\mathcal{L}\{ay\} = a\,\mathcal{L}\{y\}$, where a is a scalar constant. (From this, it follows that $\mathcal{L}\{0\} = 0$.)
2. $\mathcal{L}\{y + z\} = \mathcal{L}\{y\} + \mathcal{L}\{z\}$.

For example, differentiation is a linear operation.

Because linear problems are more tractable than nonlinear problems, considerable attention is devoted in applied mathematics to their solution. This relative simplicity is related to the property of superposition of solutions. A linear problem is homogeneous if $F = 0$:

$$\mathcal{L}\{y\} = 0. \qquad (3.2)$$

This is also said to be the homogeneous equation corresponding to Eq. 3.1. It follows from the properties of a linear operator that if y and z are each a solution to a linear, homogeneous problem, then $Ay + Bz$ is also a solution, where A and B are arbitrary constants. For example, $y = x$ and $y = 1$ are both solutions to the equation

$$\frac{d^2 y}{dx^2} = 0. \qquad (3.3)$$

The Newman Lectures on Mathematics
John Newman and Vincent Battaglia
Copyright © 2018 Pan Stanford Publishing Pte. Ltd.
ISBN 978-981-4774-25-3 (Hardcover), 978-1-315-10885-8 (eBook)
www.panstanford.com

Hence, the general solution to this equation is

$$y = Ax + B. \qquad (3.4)$$

We know this to be the general solutions because two independent constants are appropriate to a second-order equation. The principle of superposition can then be applied to the linear problem 1 because the general solution to that problem can be expressed as

$$y = y_p + y_h, \qquad (3.5)$$

where y_p is any particular solution to Eq. 3.1 and y_h is the general solution to the corresponding homogeneous Eq. 3.2. Thus, the original problem can be decomposed into two simpler problems. This explains the terminology used to describe the solution given by Eq. 2.5. The concept of superposition of solutions to linear problems will be used repeatedly in this book.

Problems

3.1 Is $\mathcal{L}$ a linear operator if it is defined as

   a. $\mathcal{L}\{y\} = y^2$?
   b. $\mathcal{L}\{y\} = y + 2$?
   c. $\mathcal{L}\{y\} = dy/dx$?
   d. $\mathcal{L}\{y\} = a_1(x)y$?
   e. $\mathcal{L}\{y\} = (d/dx)[a_1(x)y]$?

3.2 If $\mathcal{L}$ is a linear operator, which of the following problems are linear problems?

   a. $\mathcal{L}\{y\} = 3y$
   b. $\mathcal{L}\{y\} = 2x + b$
   c. $\mathcal{L}\{y\} = e^y$

Chapter 4

Linearization of Nonlinear Problems

The relative ease of treating linear problems frequently leads one to introduce approximations that produce a linear problem. There is something of an art in the formulation of mathematical models since the model is useless if it is intractable and equally useless if it fails to describe the salient features of the physical system. One recommended procedure would be to formulate a detailed and relatively precise model into which one can subsequently introduce approximations of a mathematical nature. It is usually possible to use the approximate solution thereby obtained to assess the validity of the approximations that have been made. Thus, the contemplation of an approximate solution is an important part of analysis.

If an approximate solution is slightly in error or is significantly in error, but only in a restricted domain, it may be possible to make a correction. This leads to the very important perturbation methods, which give a sound mathematical basis to many approximate solutions. One can state, as a general principle, that if the nature of the approximations is well understood, it should always be possible to use the approximate solution as a basis for a perturbation expansion.

While perturbation methods are strictly outside the scope of this book, we shall, from time to time, draw attention to cases where an approximate solution might be examined in detail and a perturbation expansion would be appropriate.

The Newman Lectures on Mathematics
John Newman and Vincent Battaglia
Copyright © 2018 Pan Stanford Publishing Pte. Ltd.
ISBN 978-981-4774-25-3 (Hardcover), 978-1-315-10885-8 (eBook)
www.panstanford.com

Linearization of nonlinear problems finds widespread use. In the examination of the stability of Poiseuille flow in a tube, it is only necessary to examine whether a small, arbitrary disturbance superposed on the basic, steady velocity profile will grow or decay in time or distance down the tube. In process dynamics and control, the response of a system at a given steady state to minor fluctuations in input variables or external conditions can be analyzed by linearization. In electronic amplification, vacuum tubes and transistors can be treated as linear elements, and the useful range of the equipment is thus defined.

Linearization is also used widely in the numerical solution to nonlinear problems where, by iteration or successive approximations, it is frequently possible to obtain the desired solution to the original problem. A simple example is the Newton–Raphson method for determining the root of a function $y(x) = 0$. This is illustrated graphically in Fig. 4.1. For an initial value x_0, one calculates y_0 and the derivative dy/dx. By linearization, one next calculates a second approximation:

$$y = y_0 + \frac{dy}{dx}(x_1 - x_0) = 0. \tag{4.1}$$

$$x_1 = x_0 - y_0/(dy/dx). \tag{4.2}$$

When this method works, it converges very rapidly. Many successive approximation methods are generalizations of this concept (see, for example, Chapter 12).

Let us illustrate how a particular problem might be linearized to yield useful results. An experimental flow loop is sketched in Fig. 4.2. The hydraulic characteristics of the experimental apparatus are approximated by the resistance of an orifice. The problem is that the pump introduces pulsations in the flow rate. A closed air cavity has been installed between the pump and the experimental apparatus in order to damp these pulsations. On an intuitive basis, we might anticipate that low-frequency pulsations would be little damped, while high-frequency pulsations might be effectively eliminated. How should the air cavity be designed without the solution to a complicated nonlinear problem? We shall neglect inertial effects here, although they can alter significantly the damping characteristics. We shall also neglect the hydrostatic head in the air cavity and take $p_c = p_1$ (see Fig. 4.2).

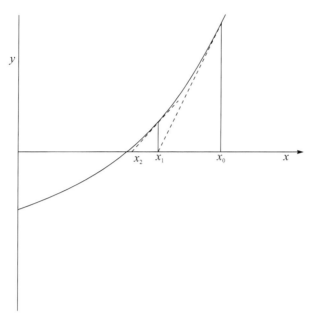

Figure 4.1 Use of the Newton–Raphson method to find x such that $y(x) = 0$.

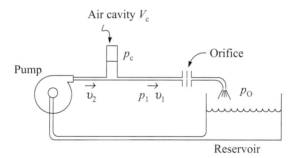

Figure 4.2 The use of an air cavity to damp pulsations in a flow loop.

Let the various flow quantities be represented as sums of their average values and oscillating parts, presumed to be small:

$$\left.\begin{aligned} v_1 &= v + v_1' \\ v_2 &= v + v_2' \\ p_c &= p + p' \\ V_c &= V + V' \end{aligned}\right\}. \qquad (4.3)$$

The resistance of the orifice is described by the equation

$$p_c - p_0 = \frac{\rho}{2C^2}\left(\frac{A_T^2}{A_0^2} - 1\right)v_1 \mid v_1 \mid, \tag{4.4}$$

where A_0 is the area of the orifice opening, A_T is the cross-sectional area of the tubing, and C is the orifice coefficient. The pressure in the air cavity is described by an equation for adiabatic expansion

$$p_c V_c^\gamma = \text{constant}, \tag{4.5}$$

where γ is the polytropic coefficient of the gas. The material balance on the liquid is

$$\frac{dV_c}{dt} = A_T(v_1 - v_2). \tag{4.6}$$

With Eq. 4.3, Eq. 4.4 for the orifice becomes

$$p - p_0 + p' = \frac{\rho}{2C^2}\left(\frac{A_T^2}{A_0^2} - 1\right)\left(v^2 + 2vv_1' + (v_1')^2\right). \tag{4.7}$$

The square of the small term v_1' is to be neglected in the linearization. Then the steady components of Eq. 4.7 can be equated

$$p - p_0 = \frac{\rho}{2C^2}\left(\frac{A_T^2}{A_0^2} - 1\right)v^2, \tag{4.8}$$

and the nonsteady components can be equated

$$p' = \frac{\rho}{2C^2}\left(\frac{A_T^2}{A_0^2} - 1\right)2vv_1'. \tag{4.9}$$

Equation 4.5 can be linearized by differentiation

$$V_c^\gamma \frac{dp'}{dt} + \gamma p_c V_c^{\gamma-1}\frac{dV'}{dt} = 0. \tag{4.10}$$

When the squares of the small, oscillating terms are neglected, this becomes

$$\frac{dV'}{dt} = -\frac{V}{\gamma p}\frac{dp'}{dt}. \tag{4.11}$$

Equation 4.6 is already linear and can be written as

$$\frac{dV'}{dt} = A_T(v_1' - v_2'). \tag{4.12}$$

Combination of Eqs. 4.9, 4.11, and 4.12 gives

$$\frac{dV'}{dt} = A_T(v_1' - v_2') = -\frac{V}{\gamma p}\frac{dp'}{dt} = -\frac{V}{\gamma p}\frac{\rho}{C^2}\left(\frac{A_T^2}{A_0^2} - 1\right)v\frac{dv_1'}{dt} \quad (4.13)$$

or

$$\tau\frac{dv_1'}{dt} + v_1' = v_2', \quad (4.14)$$

where

$$\tau = \frac{V\rho v}{\gamma p A_T C^2}\left(\frac{A_T^2}{A_0^2} - 1\right). \quad (4.15)$$

Suppose now that the pump output can be represented as

$$v_2' = A \sin \omega t. \quad (4.16)$$

After transients have decayed, the oscillating velocity through the orifice will be

$$v_1' = B \sin (\omega t + \phi). \quad (4.17)$$

The ratio of the amplitude of the oscillation of the velocity in the orifice to that in the velocity of the pump output will be

$$\frac{|v_1'|}{|v_2'|} = \frac{B}{A} = \frac{1}{\sqrt{1 + \omega^2\tau^2}}. \quad (4.18)$$

We see from these results that high-frequency oscillations will be more strongly damped than low-frequency oscillations. Furthermore, better damping occurs for a large volume of the air cavity and a high resistance of the orifice and will also depend on the flow velocity.

Chapter 5

Reduction of Order

If, in a differential equation, either the independent or the dependent variable does not appear, it is possible to reduce the order of the equation. For example, the variable y is not explicitly present in the equation

$$\frac{d^2 y}{dx^2} + 3x^2 \frac{dy}{dx} = 0. \tag{5.1}$$

Introduction of the variable

$$p = \frac{dy}{dx} \tag{5.2}$$

reduces the equation to

$$\frac{dp}{dx} + 3x^2 p = 0. \tag{5.3}$$

This linear, first-order equation with separable variables has the solution

$$p = \frac{dy}{dx} = Ae^{-x^3}. \tag{5.4}$$

A second integration yields the general solution to Eq. 5.1:

$$y = A \int_0^x e^{-x^3} dx + B. \tag{5.5}$$

The Newman Lectures on Mathematics
John Newman and Vincent Battaglia
Copyright © 2018 Pan Stanford Publishing Pte. Ltd.
ISBN 978-981-4774-25-3 (Hardcover), 978-1-315-10885-8 (eBook)
www.panstanford.com

The equation

$$\frac{d^2T}{dx^2} = K(T - T_a) \tag{5.6}$$

with the boundary conditions

$$\left.\begin{array}{l} T = T_W \text{ at } x = 0 \\ dT/dx = 0 \text{ at } x = L \end{array}\right\} \tag{5.7}$$

describes the temperature distribution in a cooling fin [1] whose thickness is small compared to its height L. Here T_W is the temperature of the wall at the base of the fin, T_a is the temperature of the air flowing past the fin, and K is a constant involving the heat-transfer coefficient between the fin and the air, the thermal conductivity of the fin, and the thickness of the fin. Since the independent variable x does not appear explicitly in the differential equation, the order of the equation can be reduced from two to one. Since Eq. 5.6 is a linear equation with constant coefficients, its solution can be effected by standard techniques (see Chapter 7). Its solution here will be relegated to the problems.

In the treatment of porous electrodes [2], the following nonlinear problem arises:

$$\frac{d^2y}{dx^2} = \frac{dy}{dx}(\delta y - \varepsilon), \tag{5.8}$$

with the boundary conditions

$$y = 0 \text{ at } x = 0 \text{ and } y = 1 \text{ at } x = 1. \tag{5.9}$$

Here δ and ε are constants. Here again, the independent variable x does not appear. Introduce

$$p = \frac{dy}{dx}, \text{ so that } \frac{d^2y}{dx^2} = \frac{dp}{dx} = \frac{dp}{dy}\frac{dy}{dx} = p\frac{dp}{dy}. \tag{5.10}$$

Equation 5.8 becomes

$$dp/dy = \delta y - \varepsilon \tag{5.11}$$

with the solution

$$p = \frac{dy}{dx} = \frac{1}{2}\delta y^2 - \varepsilon y + A. \tag{5.12}$$

Note that here, where x is absent from the equation, p is expressed in terms of y. (In contrast, in the first example where y was absent from the differential equation, p was expressed in Eq. 5.4 in terms of x.) Integration of Eq. 5.12 gives

$$\frac{1}{\theta}\tan^{-1}\left(\frac{\delta y-\varepsilon}{2\theta}\right)=x-\frac{\psi}{\theta}, \qquad (5.13)$$

where $\theta=\frac{1}{2}\sqrt{2\delta A-\varepsilon^2}$ and ψ/θ is a constant of integration. Rearrangement of Eq. 5.13 gives the general solution to Eq. 5.8:

$$y=\frac{2\theta}{\delta}\tan(\theta x-\psi)+\frac{\varepsilon}{\delta}. \qquad (5.14)$$

Since Eq. 5.8 is nonlinear, Eq. 5.14 does not represent the superposition of two linearly independent solutions, and the integration constants θ and ψ are not in positions that permit their easy evaluation. To satisfy the boundary conditions 9, θ and ψ must be calculated by trial and error from the equations

$$\tan\theta=\frac{2\theta\delta}{4\theta^2-\varepsilon(\delta-\varepsilon)} \quad\text{and}\quad \tan\psi=\frac{\varepsilon}{2\theta}. \qquad (5.15)$$

The physical situation giving rise to this problem requires that y be finite between $x = 0$ and $x = 1$. This means that $-\psi$ and $\theta-\psi$ must lie on the same branch of the tangent function, which in turn requires that $0 < \theta < \pi$.

References

1. R. Byron Bird, Warren E. Stewart, and Edwin N. Lightfoot. *Transport Phenomena*, revised 2nd ed. New York: Wiley (2007).
2. John S. Newman and Charles W. Tobias. "Theoretical analysis of current distribution in porous electrodes." *Journal of the Electrochemical Society*, **109**, 1183–1191 (1962).

Problems

5.1 Derive, by the method of this chapter, the solution to Eqs. 5.6 and 5.7.

5.2 In the calculation of the potential and charge distribution near a plane boundary of an ionized gas or an electrolytic solution, one encounters the nonlinear problem

$$\frac{d^2\phi}{dx^2} = \sinh\phi, \quad \phi = \phi_0 \text{ at } x = 0, \quad \phi \to 0 \text{ as } x \to \infty.$$

Solve for the potential ϕ.

Chapter 6

Linear, Second-Order Differential Equations

A linear, second-order equation can be written in the general form

$$\frac{d^2 y}{dx^2} + a_1(x)\frac{dy}{dx} + a_2(x)y = f(x).\tag{6.1}$$

It is not possible to write down immediately the general solution. However, one can proceed to that goal if one knows a solution $y_1(x)$ of the corresponding homogeneous equation, that is, if y_1 satisfies the equation

$$\frac{d^2 y_1}{dx^2} + a_1(x)\frac{dy_1}{dx} + a_2(x)y_1 = 0.\tag{6.2}$$

With this knowledge, let us seek the general solution by defining a new dependent variable z:

$$y = zy_1.\tag{6.3}$$

The differential equation for z now becomes

$$\frac{d^2 z}{dx^2} + \left(a_1 + 2\frac{d\ln y_1}{dx}\right)\frac{dz}{dx} = \frac{f(x)}{y_1(x)}.\tag{6.4}$$

Since z no longer appears in the equation, we can apply reduction in order techniques (see Chapter 5). The equation for $p = dz/dx$ is

$$\frac{dp}{dx} + \left(a_1 + 2\frac{d\ln y_1}{dx}\right)p = \frac{f(x)}{y_1(x)}.\tag{6.5}$$

The Newman Lectures on Mathematics
John Newman and Vincent Battaglia
Copyright © 2018 Pan Stanford Publishing Pte. Ltd.
ISBN 978-981-4774-25-3 (Hardcover), 978-1-315-10885-8 (eBook)
www.panstanford.com

This is a linear, first-order equation and has the solution (see Chapter 2)

$$p = \frac{dz}{dx} = \frac{1}{y_1^2 G} \int_{L_2}^x y_1 fG\,dx + \frac{A}{y_1^2 G}, \tag{6.6}$$

where

$$G = e^{\int_{L_1}^x a_1 dx}. \tag{6.7}$$

If Eq. 6.6 is integrated again and the result is expressed in terms of y, then the general solution to Eq. 6.1 is found to be

$$y = y_1 \int_{L_3}^x \frac{1}{y_1^2 G} \int_{L_2}^x y_1 fG\,dx\,dx + Ay_1 \int_{L_4}^x \frac{dx}{y_1^2 G} + By_1, \tag{6.8}$$

where A and B are arbitrary integration constants, and the lower limits L_1, L_2, L_3, and L_4 can be chosen so as to make the expressions as simple as possible.

If we write

$$y_2 = y_1 \int_{L_4}^x \frac{dx}{y_1^2 G}, \tag{6.9}$$

then we recognize y_1 and y_2 as two linearly independent solutions to the homogeneous equation corresponding to Eq. 6.1, and the first term on the right in Eq. 6.8 is a particular solution to Eq. 6.1. This term can be integrated by parts, and the solution 6.8 rewritten as

$$y = \int_{L_2}^x \left[y_1(\tau) y_2(x) - y_1(x) y_2(\tau) \right] f(\tau) G(\tau)\,d\tau + Ay_2 + B' y_1. \tag{6.10}$$

As an example, let us solve the different equation

$$\frac{d^2 y}{dx^2} + 2x\frac{dy}{dx} - 2y = 0 \tag{6.11}$$

subject to the boundary conditions

$$\left. \begin{array}{l} y \to 0 \ \text{as}\ x \to \infty \\ y \to x \ \text{as}\ x \to -\infty \end{array} \right\}. \tag{6.12}$$

We can see by inspection that $y_1 = x$ is a solution to this homogeneous equation. Hence, with $y = xz$, the equation for z becomes

$$x\frac{d^2 z}{dx^2} + 2(1 + x^2)\frac{dz}{dx} = 0. \tag{6.13}$$

The equation for $p = dz/dx$ is

$$x\frac{dp}{dx} + 2(1+x^2)p = 0,$$ (6.14)

for which the integrating factor is xe^{x^2}.

$$x^2 e^{x^2}\frac{dp}{dx} + 2(x+x^3)e^{x^2}p = \frac{dx^2 e^{x^2}p}{dx} = 0,$$ (6.15)

with the solution

$$p = \frac{dz}{dx} = A\frac{e^{-x^2}}{x^2}.$$ (6.16)

Further integration gives

$$z = A\int^x \frac{e^{-x^2}}{x^2}dx + B$$

$$= -A\left[\frac{e^{-x^2}}{x} + 2\int_0^x e^{-x^2}dx\right] + B.$$ (6.17)

Hence, the general solution for y is

$$y = Bx - A\left[e^{-x^2} + 2x\int_0^x e^{-x^2}dx\right].$$ (6.18)

Evaluation of the integration constants by means of the boundary conditions yields the final solution:

$$y = \frac{1}{2}x - \frac{e^{-x^2}}{2\sqrt{\pi}} - \frac{x}{2}\text{erf}(x),$$ (6.19)

where

$$\text{erf}(x) = \frac{2}{\sqrt{\pi}}\int_0^x e^{-x^2}dx$$ (6.20)

is known as the error function and is tabulated [1].

Reference

1. Milton Abramowitz and Irene A. Stegun (Eds.) *Handbook of Mathematical Functions*. Washington: National Bureau of Standards, 1964.

Chapter 7

Euler's Equation and Equations with Constant Coefficients

7.1 Linear Equations with Constant Coefficients

A general homogeneous equation of this type is

$$\frac{d^n y}{dx^n} + A_1 \frac{d^{n-1} y}{dx^{n-1}} + \cdots + A_n y = 0. \tag{7.1}$$

For a nonhomogeneous version, add the term $f(x)$ to the right. A solution to the homogeneous equation is

$$y = e^{rx}, \tag{7.2}$$

where r is a constant. We can verify this by substitution:

$$r^n e^{rx} + A_1 r^{n-1} e^{rx} + \cdots + A_n e^{rx} = 0. \tag{7.3}$$

We get an algebraic equation for r:

$$r^n + A_1 r^{n-1} + \cdots + A_n = 0. \tag{7.4}$$

This should give n roots:

$$r = r_i, \, i = 1, \cdots, n. \tag{7.5}$$

If the roots are distinct, then

$$y_h = c_1 e^{r_1 x} + c_2 e^{r_2 x} + \cdots + c_n e^{r_n x} \tag{7.6}$$

The Newman Lectures on Mathematics
John Newman and Vincent Battaglia
Copyright © 2018 Pan Stanford Publishing Pte. Ltd.
ISBN 978-981-4774-25-3 (Hardcover), 978-1-315-10885-8 (eBook)
www.panstanford.com

or

$$y_h = \sum_{i=1}^{n} c_i e^{r_i x}.$$ (7.7)

Suppose there are two equal roots. For example,

$$\frac{d^2 y}{dx^2} - 2\frac{dy}{dx} + y = 0.$$ (7.8)

Then we have

$$r^2 - 2r + 1 = 0,$$ (7.9)

or

$$r_1 = 1 \text{ and } r_2 = 1.$$ (7.10)

$y_1 = e^x$ is still a solution, but we need a second, linearly independent solution. Several approaches are available for resolving this difficulty of getting y_2.

1. You have y_1. Use reduction in order to get y_2.
2. Suppose that the roots are nearly equal. Try

$$y = C_1 e^x + C_2 e^{(1+\varepsilon)x},$$ (7.11)

and look for a limit as ε becomes small.
3. Use the method of Laplace transforms, which we develop in the next chapter.

The net result is that, when we have a repeated root, we multiply by x. If r_i is repeated m times as a root, we have the corresponding m linearly independent homogeneous solutions

$$y_1 = e^{r_i x}, \quad y_2 = x e^{r_i x}, \quad ..., \quad y_m = x^{m-1} e^{r_i x}.$$ (7.12)

You can try it yourself.

Now let us consider a nonhomogeneous equation, one where $f(x) \neq 0$. Here are some possible approaches:

1. There is a class of functions $x^k e^{Rx}$ that produce members of the class when operated on by the linear operator of the differential equation with constant coefficients. (We want to restrict ourselves to integer values of $k \geq 0$.) If $f(x)$ is the sum of such terms, one can write down a particular solution by inspection. One can treat as a group all terms with a given value of R. If k is the highest power of x in this set, then one needs

$$y_p = (P_k x^k + P_{k-1} x^{k-1} + \cdots + P_0)e^{Rx} \qquad (7.13)$$

if R is not one of the roots of the operator. If R is a root repeated m times, you need to multiply the above particular solution by x^{m+1}. Note that sines and cosines are members of this class.

2. The Laplace transform is a powerful way to approach such a problem.

3. Shudder at the prospect of applying reduction in order n times. Since you have the solution to the homogeneous equation, this should be straightforward, but tedious. I believe that this is codified under the title "variation of parameters."

7.2 Equidimensional Equations

An exactly parallel development can be carried out for the Euler–Cauchy equation

$$x^n \frac{d^n y}{dx^n} + A_1 x^{n-1} \frac{d^{n-1} y}{dx^{n-1}} + \cdots + A_n y = 0. \qquad (7.14)$$

This is also called the equidimensional equation because each term is multiplied by x once for each derivative. That is, A_1, ..., A_n are dimensionless (or have the same dimensions). A solution to the homogeneous equation is

$$y = x^s. \qquad (7.15)$$

This equation transforms into an equation with constant coefficients by letting

$$X = \ln x \text{ or } x = e^X. \qquad (7.16)$$

That is why we can be sure that a parallel development is possible. However, the equations for s and r are not the same.

Here is an example.

$$x^2 \frac{d^2 y}{dx^2} - 2x \frac{dy}{dx} + y = 0. \qquad (7.17)$$

Try

$$y = x^s. \qquad (7.18)$$

Substitution gives

$$x^2 s(s-1)x^{s-2} - 2xsx^{s-1} + x^s = 0. \qquad (7.19)$$

Cancel x^s.

$$s^2 - s - 2s + 1 = 0 \qquad (7.20)$$

or

$$s^2 - 3s + 1 = 0 \qquad (7.21)$$

or

$$s = \frac{3 \pm \sqrt{9-4}}{2} = \frac{3 \pm \sqrt{5}}{2}. \qquad (7.22)$$

Thus,

$$y_h = c_1 x^{s_1} + c_2 x^{s_2}. \qquad (7.23)$$

Even though the coefficients are the same as in the example for constant coefficients, the roots are not the same, because of differences in differentiating e^{rx} and x^s.

For repeated roots, we have the same problem as with equations with constant coefficients. For

$$x^2 \frac{d^2 y}{dx^2} - x \frac{dy}{dx} + y = 0, \qquad (7.24)$$

we get $s_1 = s_2 = 1$. Thus, the solution is

$$y_h = c_1 x + c_2 x \ln x. \qquad (7.25)$$

Also for the nonhomogeneous equation, the family of functions includes

$$(\ln x)^k x^s. \qquad (7.26)$$

If the nonhomogeneous term is not a sum of terms from this family, we may need to use reduction in order or the method of variation of parameters. The Laplace transformation is not of much use, unless we transform first to an equation with constant coefficients.

Problem

7.1 Discuss how you would approach solving the nonhomogeneous linear differential equation:

$$x^2 \frac{d^2 y}{dx^2} - x \frac{dy}{dx} + y = \sin x. \qquad (7.27)$$

Chapter 8

Series Solutions and Singular Points

Frequently, it is possible to obtain a solution to a differential equation in the form of a power series:

$$y = \sum_{k=0}^{\infty} A_k (x - x_0)^k. \tag{8.1}$$

Substitution into the differential equation would then yield relationships among the coefficients of the power series. We shall restrict ourselves here to linear homogeneous equations.

For example, let us treat the equation

$$\frac{d^2 y}{dx^2} = y \tag{8.2}$$

with an expansion around $x = 0$. Direct substitution gives

$$\sum_{k=0}^{\infty} k(k-1) A_k x^{k-2} = \sum_{k=0}^{\infty} A_k x^k. \tag{8.3}$$

By changing the index of the first series to $l = k - 2$, we obtain

$$\sum_{l=-2}^{\infty} (l+2)(l+1) A_{l+2} x^l - \sum_{k=0}^{\infty} A_k x^k = 0. \tag{8.4}$$

Since the designation of the index is immaterial, we can write this as

The Newman Lectures on Mathematics
John Newman and Vincent Battaglia
Copyright © 2018 Pan Stanford Publishing Pte. Ltd.
ISBN 978-981-4774-25-3 (Hardcover), 978-1-315-10885-8 (eBook)
www.panstanford.com

$$(-2+2)(-2+1)A_0x^{-2} + (-1+2)(-1+1)A_1x^{-1}$$

$$+\sum_{k=0}^{\infty}\left[(k+2)(k+1)A_{k+2} - A_k\right]x^k = 0. \tag{8.5}$$

This equation can be satisfied for arbitrary values of x only by setting the coefficient of each power of x equal to zero. We see that this leaves A_0 and A_1 undetermined and leads to the recursion relation for higher-order coefficients:

$$(k+2)(k+1)A_{k+2} = A_k, \tag{8.6}$$

or

$$\left.\begin{array}{ll} A_k = A_0 / k! & \text{for } k \text{ even} \\ A_k = A_1 / k! & \text{for } k \text{ odd} \end{array}\right\}. \tag{8.7}$$

We can construct two linearly independent solutions to Eq. 8.2 by setting for the first $A_0 = A_1 = 1$ giving

$$y_1 = \sum_{k=0}^{\infty} \frac{x^k}{k!} = e^x \tag{8.8}$$

and for the second $A_0 = -A_1 = 1$ giving

$$y_2 = \sum_{k=0}^{\infty} \frac{(-1)^k x^k}{k!} = e^{-x}. \tag{8.9}$$

The general solution to Eq. 8.2 is thus

$$y = Ae^x + Be^{-x}, \tag{8.10}$$

where A and B are arbitrary constants. We notice furthermore that in this case, the power series converge for all values of x.

We may note, however, that all differential equations do not allow power series expansions at all points $x = x_0$. For example, the equidimensional or Euler–Cauchy equation

$$2x\frac{dy}{dx} = y \tag{8.11}$$

has the solution

$$y = Ax^{1/2}, \tag{8.12}$$

which cannot be expanded in a power series about $x = 0$.

If a linear, homogeneous equation is put in standard form by dividing through by the coefficient of the highest derivative, that is,

$$\frac{d^2 y}{dx^2} + a_1(x)\frac{dy}{dx} + a_2(x)y = 0 \qquad (8.13)$$

for a second-order equation, then its solutions can be expanded in power series (like Eq. 8.1) about $x = x_0$ if and only if the coefficients of the lower-order derivatives in Eq. 8.13 can be expanded in power series about $x = x_0$. Such a point is called an ordinary point of the differential equation. Otherwise, the point $x = x_0$ is called a singular point. Thus, Legendre's equation

$$(1-x^2)\frac{d^2 y}{dx^2} - 2x\frac{dy}{dx} + p(p+1)y = 0 , \qquad (8.14)$$

where p is a constant, has singular points only at $x = 1$ and $x = -1$.

Substitution of a power series about $x = 0$ into Legendre's equation shows that A_0 and A_1 are arbitrary, while the higher-order coefficients are determined from

$$(k + 2)(k + 1)A_{k+2} = [k^2 + k - p(p + 1)]A_k, \, k = 0, 1, \ldots \quad (8.15)$$

$$= (k + p + 1)(k - p)A_k .$$

Selection of $A_0 = 1$, $A_1 = 0$ or $A_0 = 0$, $A_1 = 1$ leads to two linearly independent solutions

$$y_1 = 1 - \frac{p(p+1)}{2!}x^2 + \frac{p(p-2)(p+1)(p+3)}{4!}x^4 + \cdots . \quad (8.16)$$

$$y_2 = x - \frac{(p-1)(p+2)}{3!}x^3 + \frac{(p-1)(p-3)(p+2)(p+4)}{5!}x^5 + \cdots .$$

$$(8.17)$$

We may notice from Eq. 8.15 that neither series converges for $|x| \geq 1$ (unless it happens to terminate). This is a general characteristic of power series solutions: in general, they converge only in a circular region in the complex plane. The circle is centered at the point $x = x_0$ about which the expansion is made and extends to the nearest singular point (which may be complex).

We may further notice that one of the series solutions 8.16 or 8.17 will terminate if p is a positive integer or zero. In this case, $p = n$, the terminating power series is used to define the so-called Legendre polynomial $P_n(x)$, constructed so that $P_n(1) = 1$:

$$P_n(x) = y_1(x)/y_1(1) \quad \text{for } n \text{ even}$$
$$P_n(x) = y_2(x)/y_2(1) \quad \text{for } n \text{ odd}$$

$$(8.18)$$

The second, linearly independent solution is in standard form denoted $Q_n(x)$:

$$Q_n(x) = y_1(1)y_2(x) \quad \text{for } n \text{ even}$$
$$Q_n(x) = -y_2(1)y_1(x) \quad \text{for } n \text{ odd}$$

$$(8.19)$$

For nonintegral values of p, it is possible to have a linear combination of the solutions $y_1(x)$ and $y_2(x)$ such that it has the finite value of 1 at $x = 1$. In standard form, this solution is denoted $P_p(x)$, and there is another form for $Q_p(x)$. Since $P_p(-1)$ is infinite unless p is integral and $Q_p(1)$ is always infinite, it follows that the only Legendre functions that are finite at both $x = 1$ and $x = -1$ are the Legendre polynomials $P_p(x)$ for which p is integral. As we shall see, this fact has important consequences in the solution to partial differential equations.

Now let us return to the problem of singular points and treat the equation

$$4x^2 \frac{d^2 y}{dx^2} + 4x \frac{dy}{dx} + (x^2 - 1)y = 0,$$

$$(8.20)$$

for which $x = 0$ is a singular point. We can find the behavior near the singular point by looking at the equation obtained as x becomes small:

$$4x^2 \frac{d^2 y}{dx^2} + 4x \frac{dy}{dx} - y = 0.$$

$$(8.21)$$

We have no *a priori* justification for dropping any of these terms. Since this is an Euler equation, we find that it has the linearly independent solutions

$$y_1 = x^{1/2} \text{ and } y_2 = x^{-1/2}.$$

$$(8.22)$$

We can go back now and verify that for either of these solutions, the term neglected in Eq. 8.20 is not important near $x = 0$ in the sense that it is small compared to any of the terms retained.

Having now determined the behavior of the solutions near $x = 0$, we should be able to go back and obtain suitable correction terms to account for the term that had been neglected. Substitution of the assumed form

$$u = \sum_{k=0}^{\infty} A_k x^{k+s} \tag{8.23}$$

where $s = 1/2$ or $-1/2$ leads to the equation

$$\sum_{k=0}^{\infty} \left[4(k+s)^2 - 1 \right] A_k x^{k+s} + \sum_{k=2}^{\infty} A_{k-2} x^{k+s} = 0. \tag{8.24}$$

This can be satisfied by setting $A_0 = 1$, $A_1 = 0$, and

$$[4(k+s)^2 - 1] A_k = -A_{k-2}, \ k = 2, 3, 4, \dots. \tag{8.25}$$

We should now generalize these concepts and bring out some additional points. If $x = x_0$ is a singular point of a linear, homogeneous differential equation, write the equation in a form illustrated for second-order equations:

$$\frac{d^2 y}{dx^2} + \frac{b_1(x)}{x - x_0} \frac{dy}{dx} + \frac{b_2(x)}{(x - x_0)^2} y = 0. \tag{8.26}$$

If b_1 and b_2 can be expanded in a power series about $x = x_0$, this point is said to be a regular singular point of the differential equation. Otherwise, it is called an irregular singular point. The former are tractable, while the latter may not be.

If, for a regular singular point, $b_1(x)$ and $b_2(x)$ are replaced by $b_1(x_0)$ and $b_2(x_0)$, an Euler equation is obtained:

$$\frac{d^2 y}{dx^2} + \frac{b_1(x_0)}{x - x_0} \frac{dy}{dx} + \frac{b_2(x_0)}{(x - x_0)^2} y = 0. \tag{8.27}$$

The behavior of the solutions near the singular point can thus be obtained by solving this equation, that is, by assuming a form $y = (x - x_0)^s$. The solutions to the resulting equation

$$s(s-1) + b_1(x_0)s + b_2(x_0) = 0 \tag{8.28}$$

for s, denoted in this case s_1 and s_2, are called the exponents of the equation at the point $x = x_0$. For an ordinary point $b_1(x_0) = b_2(x_0) = 0$, and the solutions are $s_1 = 1$ and $s_2 = 0$.

For a regular singular point, one can extend at least one of the solutions obtained above by assuming a solution of the form

$$u = \sum_{k=0}^{\infty} A_k (x - x_0)^{k+s}, \tag{8.29}$$

where s is an exponent of the equation at the singular point and $A_0 \neq$ 0. Frequently, this procedure will lead to a suitable solution for each value of s. An exception will arise if two of the exponents are equal and may arise if two exponents differ by an integer, although it did not arise in the case of Eq. 8.20 where the exponents differ by 1.

For an exceptional case, an assumed solution of the form 8.29 will always work for the larger exponent, say s_1. Let us denote this solution by u_1. Then a second solution can always be obtained in the form

$$u_2 = Cu_1(x)\ln(x-x_0) + \sum_{k=0}^{\infty} B_k (x-x_0)^{k+s_2}. \qquad (8.30)$$

The series in Eqs. 8.29 and 8.30 will always converge in a circular domain extending out to the nearest adjacent singular point.

For an irregular singular point, textbooks do not offer us much hope. Perhaps we should distinguish the case where in Eq. 8.26 $b_1(x_0)$ or $b_2(x_0)$ is infinite from the case where $b_1(x_0)$ and $b_2(x_0)$ are finite. In the latter case, we can validly reduce the equation to an Euler equation at $x = x_0$, and it should be possible to use the Euler solution as a basis for obtaining corrections, even though a solution of the form 8.29 will not result.

This section on series solutions may seem tedious. However, careful study here should prepare the student for more interesting perturbation methods, including for singular perturbations in the trilogy book on transport phenomena.

Problems

8.1 Investigate the behavior of solutions near the singular point $x = +1$ of Legendre's equation.

8.2 The differential equation

$$2x\frac{d^2y}{dx^2} + (1-x^{1/2})\frac{dy}{dx} = 0$$

has an irregular singular point at $x = 0$. Try to use the solutions to the corresponding Euler equation as a basis for obtaining a valid solution. This example was selected so that you could also obtain a solution by reduction-in-order techniques. Furthermore, the transformation $t = x^{1/2}$ yields an equation for $y(t)$ for which $t = 0$ happens to be an ordinary point.

Chapter 9

Legendre's Equation and Special Functions

We treated in the last chapter Legendre's equation

$$(1-x^2)\frac{d^2y}{dx^2} - 2x\frac{dy}{dx} + p(p+1)y = 0, \tag{9.1}$$

where p is a constant, and obtained power series solutions

$$y = \sum_{k=0}^{\infty} A_k x^k, \tag{9.2}$$

where A_0 and A_1 are arbitrary and the higher-order coefficients are related by

$$(k+2)(k+1)A_{k+2} = (k+p+1)(k-p)A_k, \, k = 0, 1, \dots . \tag{9.3}$$

We defined two linearly independent solutions as

$$y_1 = 1 - \frac{p(p+1)}{2!}x^2 + \frac{p(p-2)(p+1)(p+3)}{4!}x^4 + \cdots . \tag{9.4}$$

$$y_2 = x - \frac{(p-1)(p+2)}{3!}x^3 + \frac{(p-1)(p-3)(p+2)(p+4)}{5!}x^5 + \cdots .$$

$$\tag{9.5}$$

We noted that these series do not converge beyond the singular points $|x| \geq 1$ unless they terminate, and one or the other terminates only if p is an integer. This led us to define the Legendre polynomials:

The Newman Lectures on Mathematics
John Newman and Vincent Battaglia
Copyright © 2018 Pan Stanford Publishing Pte. Ltd.
ISBN 978-981-4774-25-3 (Hardcover), 978-1-315-10885-8 (eBook)
www.panstanford.com

$$P_n(x) = y_1(x)/y_1(1) \quad \text{for} \quad p = n, \text{an even integer}$$
$$P_n(x) = y_2(x)/y_2(1) \quad \text{for} \quad p = n, \text{an odd integer} \quad , \qquad (9.6)$$

and associated Legendre functions

$$Q_n(x) = y_1(1)y_2(x) \quad \text{for} \quad p = n, \text{an even integer}$$
$$Q_n(x) = -y_2(1)y_1(x) \quad \text{for} \quad p = n, \text{an odd integer} \quad . \qquad (9.7)$$

In this case, as in many other cases, a differential equation generates solutions that are important enough to justify defining special functions that are tabulated and whose properties are studied extensively. Frequently, though not in the case of the Legendre polynomials, these special functions cannot be expressed in terms of elementary functions. Consequently, it is important to be able to recognize when a problem has a solution that can be expressed in terms of tabulated functions with well-known properties. Some important examples, particularly Bessel's functions, are discussed in textbooks, but it does not seem appropriate to repeat a lot of detailed equations and derivations here. Such special functions should be no more frightening than sines and cosines, which can be regarded as solutions to differential equations that have been tabulated and whose properties are known. When the need for special functions arises, one should look them up in a source such as Abramowitz and Stegun.

Let us look further into the properties of Legendre functions. Some of these are plotted in Fig. 9.1. The Legendre polynomials are plotted against $\theta = \cos^{-1}x$ because in many physical problems where they arise, θ represents the azimuthal angle in spherical coordinates. We see that they do somewhat resemble the functions $\cos n\theta$.

One interesting and highly useful property of the Legendre polynomials is that they are orthogonal in the sense that

$$\int_{-1}^{1} P_n(x)P_m(x)dx = 0 \quad \text{unless } m = n. \qquad (9.8)$$

We derive this result by writing down the differential equations satisfied by the functions $P_n(x)$ and $P_m(x)$:

$$\frac{d}{dx}\left((1-x^2)\frac{dP_n}{dx} \right) = -n(n+1)P_n. \qquad (9.9)$$

$$\frac{d}{dx}\left((1-x^2)\frac{dP_m}{dx} \right) = -m(m+1)P_m. \qquad (9.10)$$

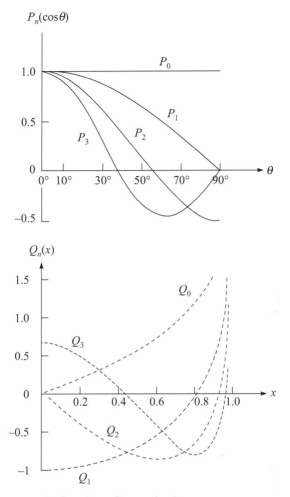

Figure 9.1 Legendre functions of integral order.

Multiply the first equation by P_m, the second by P_n, subtract and integrate from -1 to $+1$ to get

$$\int_{-1}^{1}\left[P_m\frac{d}{dx}\left((1-x^2)\frac{dP_n}{dx}\right)-P_n\frac{d}{dx}\left((1-x^2)\frac{dP_m}{dx}\right)\right]dx$$

$$=[m(m+1)-n(n+1)]\int_{-1}^{1}P_nP_mdx. \tag{9.11}$$

Integration by parts gives

$$P_m(1-x^2)\frac{dP_n}{dx}\bigg|_{-1}^{1} - \int_{-1}^{1}\frac{dP_m}{dx}(1-x^2)\frac{dP_n}{dx}dx - P_n(1-x^2)\frac{dP_m}{dx}\bigg|_{-1}^{1}$$

$$+\int_{-1}^{1}\frac{dP_n}{dx}(1-x^2)\frac{dP_m}{dx}dx = [m(m+1)-n(n+1)]\int_{-1}^{1}P_nP_mdx. \qquad (9.12)$$

The integrals cancel, and $1-x^2$ vanishes at $x = \pm1$, and we are left with the result

$$[m(m+1)-n(n+1)]\int_{-1}^{1}P_nP_mdx = 0, \qquad (9.13)$$

which corresponds to Eq. 9.8.

Since $P_n(x)$ is an even or odd function of x depending on whether n is even or odd, it further follows from the above result that the even Legendre polynomials are orthogonal over the interval from 0 to 1, and similarly for the odd Legendre polynomials:

$$\int_0^1 P_{2m}(x)P_{2n}(x)dx = 0 \quad \text{if } m \neq n. \qquad (9.14)$$

$$\int_0^1 P_{2m+1}(x)P_{2n+1}(x)dx = 0 \quad \text{if } m \neq n. \qquad (9.15)$$

However, an even Legendre polynomial is not orthogonal to an odd Legendre polynomial over the interval from 0 to 1, only over the interval from –1 to +1.

The orthogonal property of these functions is useful because one frequently wants to express an arbitrary function $f(x)$ as the sum of Legendre polynomials

$$f(x) = \sum_{k=0}^{\infty} B_k P_k(x) \quad \text{for } -1 < x < 1. \qquad (9.16)$$

The coefficients can be properly evaluated by multiplying by $P_n(x)$ and integrating from –1 to +1.

$$\int_{-1}^{+1} f(x)P_n(x)dx$$

$$= \sum_{k=0}^{\infty} B_k \int_{-1}^{+1} P_k(x)P_n(x)dx = B_n \int_{-1}^{+1}\left[P_n(x)\right]^2 dx. \qquad (9.17)$$

Only one term remains in the series on the right because the Legendre polynomials are orthogonal. Hence

$$B_n = \int_{-1}^{1} f(x)P_n(x)dx \bigg/ \int_{-1}^{1} \left[P_n(x)\right]^2 dx. \qquad (9.18)$$

Since

$$\int_{-1}^{1} \left[P_n(x)\right]^2 dx = \frac{1}{n+\dfrac{1}{2}}, \qquad (9.19)$$

Eq. 9.18 becomes

$$B_n = \frac{2n+1}{2} \int_{-1}^{1} f(x)P_n(x)dx. \qquad (9.20)$$

Sometimes it is possible to evaluate these integrals. For example,

$$\int_{0}^{1} P_n(x)x^r dx = \frac{\sqrt{\pi}}{2^{1+r}} \frac{\Gamma(1+r)}{\Gamma\left(1+\dfrac{1}{2}r-\dfrac{1}{2}n\right)\Gamma\left(\dfrac{1}{2}r+\dfrac{1}{2}n+\dfrac{3}{2}\right)}, \qquad (9.21)$$

where Γ refers to gamma function.

Such series expansions of given functions are similar to Fourier sine or cosine series and are, in fact, called Fourier–Legendre series. The Legendre polynomials are complete in the interval from −1 to +1 in the sense that any function that is piecewise continuous in that interval can be expanded in such a series 9.16 with the coefficients evaluated from Eq. 9.20, and the series so obtained will converge to $f(x)$ at all points between −1 and 1 where $f(x)$ is continuous and to $\frac{1}{2}[f(x+)+f(x-)]$ at points where $f(x)$ is discontinuous.

For an expansion that needs to be valid only in the interval from 0 to 1, it is sufficient to use either the even or the odd Legendre polynomials:

$$f(x)=\sum_{n=0}^{\infty}C_n P_{2n}(x) \quad \text{for } 0<x<1 \qquad (9.22)$$

and

$$f(x)=\sum_{n=0}^{\infty}D_n P_{2n+1}(x) \quad \text{for } 0<x<1, \qquad (9.23)$$

with the coefficients evaluated as follows:

$$C_n = (4n+1)\int_0^1 f(x)P_{2n}(x)dx$$

and

$$D_n = (4n+3)\int_0^1 f(x)P_{2n+1}(x)dx. \tag{9.24}$$

In the case of Eq. 9.23, the series will not converge to $f(x)$ at $x = 0$ unless $f(0) = 0$, since $P_{2n+1}(0) = 0$.

For the numerical calculation of Legendre polynomials, one uses the recursion formula

$$(n + 1)P_{n+1}(x) = (2n+1)xP_n(x) - nP_{n-1}(x), \tag{9.25}$$

starting with $P_0 = 1$ and $P_1 = x$. For large n, this is better than the series expansion since the terms there are of alternating signs and the coefficients become very large. Use of Eq. 9.25 gives good percentage accuracy except near the zeros of the calculated polynomial.

Finally, we might mention the formula of Rodrigues

$$P_n(x) = \frac{1}{2^n n!}\frac{d^n(x^2-1)^n}{dx^n}, \tag{9.26}$$

and a recursion formula involving the derivative

$$(x^2-1)\frac{dP_n}{dx} = nxP_n(x) - nP_{n-1}(x). \tag{9.27}$$

Chapter 10

Laplace Transformation

Here we encounter a method particularly well suited for linear differential equations with constant coefficients and where the boundary conditions are all given at one point (initial-value problems). In this chapter, we shall denote the independent variable as t, which may frequently be regarded as representing time.

Given a function $f(t)$, the Laplace transform of f with respect to t, variously denoted as $\mathscr{L}\{f(t)\}$, $\overline{f}(s)$, or $F(s)$, is defined as

$$\mathscr{L}\{f(t)\} = F(s) = \int_0^\infty f(t)\, e^{-st} dt, \tag{10.1}$$

provided that the integral exists. Obviously, f must be defined in the domain from 0 to ∞ (except on a set of measure zero) and $f(t)e^{-st}$ must approach zero as $t \to \infty$, at least for those values of s for which the transform is defined.

The value of the Laplace transformation lies in the fact that when it is applied to a linear differential equation with constant coefficients, one obtains an algebraic equation for the transform of the unknown. Let us seek the transform of the derivative df/dt of f:

$$\mathscr{L}\left\{\frac{df}{dt}\right\} = \int_0^\infty \frac{df}{dt} e^{-st} dt. \tag{10.2}$$

The Newman Lectures on Mathematics

John Newman and Vincent Battaglia

Copyright © 2018 Pan Stanford Publishing Pte. Ltd.

ISBN 978-981-4774-25-3 (Hardcover), 978-1-315-10885-8 (eBook)

www.panstanford.com

Integration by parts gives

$$\mathscr{L}\left\{\frac{df}{dt}\right\} = fe^{-st}\Big|_0^\infty + s\int_0^\infty fe^{-st}dt. \tag{10.3}$$

Since fe^{-st} vanishes at $t = \infty$ and since the integral is the Laplace transform of f, we have

$$\mathscr{L}\left\{\frac{df}{dt}\right\} = sF(s) - f(0). \tag{10.4}$$

Let us illustrate this with a few examples.

$$\mathscr{L}\left\{e^{at}\right\} = \int_0^\infty e^{at}e^{-st}dt = \frac{e^{(a-s)t}}{a-s}\Big|_0^\infty = \frac{1}{s-a}. \tag{10.5}$$

(We see that this is defined only for $s > a$.) Thus, the Laplace transform of the derivative is

$$\mathscr{L}\left\{\frac{de^{at}}{dt}\right\} = \mathscr{L}\left\{ae^{at}\right\} = a\mathscr{L}\left\{e^{at}\right\} = \frac{a}{s-a}. \tag{10.6}$$

On the other hand, from Eq. 10.4, we get

$$\mathscr{L}\left\{\frac{de^{at}}{dt}\right\} = s\mathscr{L}\left\{e^{at}\right\} - 1 = \frac{s}{s-a} - 1 = \frac{a}{s-a}, \tag{10.7}$$

which checks.

From Eq. 10.5, it follows that

$$\mathscr{L}\{1\} = \frac{1}{s}, \tag{10.8}$$

and by direct calculation

$$\mathscr{L}\{t\} = \int_0^\infty te^{-st}dt = \frac{e^{-st}}{-s}t\Big|_0^\infty - \int_0^\infty \frac{e^{-st}}{-s}dt = \frac{1}{s^2}. \tag{10.9}$$

We can check Eq. 10.4 with this example:

$$\mathscr{L}\{1\} = \mathscr{L}\left\{\frac{dt}{dt}\right\} = s\mathscr{L}\{t\} - 0 = \frac{s}{s^2} = \frac{1}{s}. \tag{10.10}$$

Now let

$$\left.\begin{array}{l} f = t \quad \text{for} \quad t < 1 \quad \text{and} \quad f = 0 \quad \text{for} \quad t > 1 \\ g = 1 \quad \text{for} \quad t < 1 \quad \text{and} \quad g = 0 \quad \text{for} \quad t > 1 \end{array}\right\}. \tag{10.11}$$

The Laplace transforms are calculated to be

$$\mathscr{L}\{f\} = \int_0^1 te^{-st}\,dt = -\frac{1}{s}e^{-s} - \frac{1}{s^2}e^{-s} + \frac{1}{s^2}. \qquad (10.12)$$

$$\mathscr{L}\{g\} = \int_0^1 e^{-st}\,dt = -\frac{1}{s}e^{-s} + \frac{1}{s}. \qquad (10.13)$$

From Eq. 10.4

$$\mathscr{L}\left\{\frac{df}{dt}\right\} = -e^{-s} - \frac{1}{s}e^{-s} + \frac{1}{s} - 0 \neq \mathscr{L}\{g\}. \qquad (10.14)$$

We are thus led to conclude that either the formula 10.4 does not work or that g is not the derivative of f. The trouble is that f is not the integral of g because if it were, then f would be continuous. Let

$$h = \int_0^t g\,dt = \begin{cases} t & \text{for } t < 1 \\ 1 & \text{for } t > 1 \end{cases}. \qquad (10.15)$$

Then

$$\mathscr{L}\{h\} = -\frac{1}{s^2}e^{-s} + \frac{1}{s^2}. \qquad (10.16)$$

The functions g and h then satisfy Eq. 10.4 with $g = dh/dt$, and we are led to conclude that Eq. 10.4 is valid only if f is continuous.

Another somewhat artificial way to make Eq. 10.4 work is to say that the derivative of f is

$$\ell = \frac{df}{dt} = g(t) - \delta(t-1), \qquad (10.17)$$

where δ satisfies the conditions

$$\left.\begin{array}{l} \delta(t) = 0 \quad \text{if } t \neq 0 \\ \displaystyle\int_{-\infty}^{+\infty} \delta(t)\,dt = 1 \end{array}\right\}. \qquad (10.18)$$

Thus, δ is really not a function, but now f is the integral of ℓ :

$$f = \int_0^t \ell(t)\,dt. \qquad (10.19)$$

Furthermore, the Laplace transform of ℓ is

$$\mathscr{L}\{\ell\} = \mathscr{L}\{g\} - \int_0^\infty \delta(t-1)e^{-st}\,dt = \mathscr{L}\{g\} - e^{-s}, \qquad (10.20)$$

and ℓ and f now satisfy Eq. 10.4 with $f = d\ell/dt$.

δ is an example of so-called "singularity functions," whose introduction is frequently convenient because they allow formal treatment of cases that, otherwise, would require special treatment.

In Chapter 5, we posed the problem for a heat-transfer fin:

$$\frac{d^2T}{dx^2} = K(T - T_a) \qquad (10.21)$$

with the boundary conditions

$$\left.\begin{array}{ll} T = T_W & \text{at} \quad x = 0 \\ dT/dx = 0 & \text{at} \quad x = L \end{array}\right\}. \qquad (10.22)$$

This is not a typical problem for which the Laplace transformation is ideally suited since the boundary conditions are not both stated at $x = 0$ and since the domain of x does not extend to infinity. Nevertheless, we can use this problem to illustrate the method.

Here the transformation is with respect to x rather than t. Denote the transform of $T(x)$ as $\overline{T}(s)$. Repeated use of Eq. 10.4 gives

$$\mathscr{L}\left\{\frac{d^2T}{dx^2}\right\} = s\mathscr{L}\left\{\frac{dT}{dx}\right\} - T'(0) = s^2\overline{T}(s) - sT(0) - T'(0). \quad (10.23)$$

We know that $T(0) = T_w$, but we do not know $T'(0)$ and must carry it as an unknown constant. Application of the Laplace transformation to Eq. 10.21 thus gives an algebraic equation for the transform $\overline{T}(s)$:

$$s^2\overline{T}(s) - sT_W - T'(0) = K\overline{T}(s) - \frac{KT_a}{s}, \qquad (10.24)$$

where Eq. 10.8 has been used to evaluate the transform of the nonhomogeneous term. Equation 10.24 has the solution

$$\overline{T}(s) = \frac{sT_W}{s^2 - K} + \frac{T'(0)}{s^2 - K} - \frac{KT_a}{s(s^2 - K)}. \qquad (10.25)$$

The inversion of Eq. 10.25 to obtain $T(x)$ can be effected by rewriting it in terms whose inverses we already know:

$$\overline{T}(s) = \frac{\frac{1}{2}\left[T_W - T_a + T'(0)/\sqrt{K}\right]}{s - \sqrt{K}} + \frac{\frac{1}{2}\left[T_W - T_a - T'(0)/\sqrt{K}\right]}{s + \sqrt{K}} + \frac{T_a}{s}.$$

(10.26)

Thus, the function $T(x)$ is

$$T(x) = \frac{1}{2}\left[T_W - T_a + T'(0)/\sqrt{K}\right]e^{\sqrt{K}x}$$
$$+ \frac{1}{2}\left[T_W - T_a - T'(0)/\sqrt{K}\right]e^{-\sqrt{K}x} + T_a.$$

(10.27)

We evaluate $T'(0)$ so that we satisfy the boundary condition $dT/dx = 0$ at $x = L$:

$$T'(0) = -\sqrt{K}(T_W - T_a)\tanh\sqrt{k}L.$$

(10.28)

The final solution for $T(x)$ thus becomes

$$T(x) = T_a + (T_W - T_a)\left[\cosh(\sqrt{K}x) - \tanh(\sqrt{K}L)\sinh(\sqrt{K}x)\right]$$

$$= T_a + (T_W - T_a)\frac{\cosh\left[\sqrt{K}(L-x)\right]}{\cosh\left(\sqrt{K}L\right)}.$$

(10.29)

The method of Laplace transforms is also quite useful for coupled, linear differential equations with constant coefficients. Let us treat the system

$$\frac{dy}{dt} + 2x = 0, \quad \frac{dx}{dt} + 2y = e^{at}, \quad x(0) = y(0) = 0.$$

(10.30)

Taking Laplace transforms, we obtain

$$s\overline{y} + 2\overline{x} = 0, \quad s\overline{x} + 2\overline{y} = \frac{1}{s-a},$$

(10.31)

which we can readily solve for the transforms themselves:

$$\overline{x} = \frac{s}{(s-a)(s^2-4)} = \frac{a}{a^2-4}\frac{1}{s-a} + \frac{1/2}{2-a}\frac{1}{s-2} - \frac{1/2}{2+a}\frac{1}{s+2}.$$

(10.32)

$$\overline{y} = \frac{-2}{(s-a)(s^2-4)} = \frac{2}{4-a^2}\frac{1}{s-a} - \frac{1/2}{2-a}\frac{1}{s-2} - \frac{1/2}{2+a}\frac{1}{s+2}.$$

(10.33)

Here, again, we have tried to use simpler terms whose inverses we know. The solutions for x and y thus are

$$x = \frac{a}{a^2 - 4}e^{at} + \frac{1/2}{2-a}e^{2t} - \frac{1/2}{2+a}e^{-2t}, \qquad (10.34)$$

$$y = \frac{2}{4-a^2}e^{at} - \frac{1/2}{2-a}e^{2t} - \frac{1/2}{2+a}e^{-2t}, \qquad (10.35)$$

as we can verify by direct substitution. Such systems can be treated by other methods that are usually more complicated.

We see that the application of Laplace transformation to linear, differential equations with constant coefficients and the solution to the transformed equations for the transforms is straightforward, and we are usually faced with the problem of obtaining the inverse of the Laplace transforms. This is facilitated by the compilation of transforms, so that with a little ingenuity, one can simply look up the required functions. Although more general methods of determining inverse transforms exist (see Chapter 20), we shall not delve into them here. Instead we shall occupy ourselves with the determination of some additional transforms, followed by some asymptotic methods and the consideration of a special linear differential equation with one variable coefficient.

Let us evaluate the Laplace transform of t^n, where $n > -1$:

$$\mathscr{L}\{t^n\} = \int_0^\infty t^n e^{-st} dt. \qquad (10.36)$$

Introduction of a new variable $x = st$ allows this to be written as

$$\mathscr{L}\{t^n\} = \frac{1}{s^{n+1}} \int_0^\infty x^n e^{-x} dx. \qquad (10.37)$$

Now the integral is independent of s and is a constant depending only on the parameter n. Thus, the dependence of the transform on the variable s is shown clearly by Eq. 10.37. The integral can be expressed as the "gamma function," which is tabulated in many places:

$$\mathscr{L}\{t^n\} = \frac{\Gamma(n+1)}{s^{n+1}}. \qquad (10.38)$$

If n is an integer, the result can be expressed with the factorial:

$$\mathscr{L}\{t^n\} = \frac{n!}{s^{n+1}}, \quad n = 0, 1, 2, \dots. \qquad (10.39)$$

One can compare this with the special cases given in Eqs. 10.8 and 10.9. Further, since $\Gamma\left(\dfrac{1}{2}\right) = \sqrt{\pi}$, we can write

$$\mathcal{L}\{1\sqrt{t}\} = \frac{\sqrt{\pi}}{\sqrt{s}}. \tag{10.40}$$

An interesting result is obtained if we differentiate the Laplace transform with respect to s:

$$\frac{dF(s)}{ds} = \frac{d}{ds}\int_0^\infty f(t)e^{-st}\,dt = -\int_0^\infty tf(t)\,e^{-st}\,dt. \tag{10.41}$$

For multiple differentiations, we thus obtain

$$\mathcal{L}\{t^n f(t)\} = (-1)^n \frac{d^n F(s)}{ds^n}. \tag{10.42}$$

Application to Eq. 10.40 gives

$$\mathcal{L}\{\sqrt{t}\} = \mathcal{L}\{t/\sqrt{t}\} = \frac{\sqrt{\pi}}{2s^{3/2}}. \tag{10.43}$$

Sometimes it is useful to be able to invert the product of the transforms of two known functions:

$$F(s)G(s) = \int_0^\infty f(t)e^{-st}\,dt \int_0^\infty g(t)e^{-st}\,dt. \tag{10.44}$$

In one of the integrals, replace the dummy variable of integration by u:

$$F(s)G(s) = \int_0^\infty f(t)e^{-st}\,dt \int_0^\infty g(u)e^{-su}\,du = \int_0^\infty\int_0^\infty f(t)g(u)e^{-(t+u)s}\,dt\,du. \tag{10.45}$$

Now let the variable t be replaced by $v = t + u$

$$F(s)G(s) = \int_0^\infty\int_u^\infty f(v-u)g(u)e^{-vs}\,dv\,du. \tag{10.46}$$

Interchanging the order of integration gives

$$F(s)G(s) = \int_0^\infty \int_0^v f(v-u)g(u)e^{-vs}\,du\,dv = \int_0^\infty e^{-vs}\int_0^v f(v-u)g(u)\,du\,dv.$$

$$(10.47)$$

If we let $v = t$ and $u = \tau$, we see that the product $F(s)G(s)$ is the transform of the inner integral, or

$$\mathscr{L}^{-1}\{F(s)G(s)\} = \int_0^t f(t-\tau)g(\tau)d\tau. \qquad (10.48)$$

From the symmetry of the product $F(s)G(s)$, it is clear that the arguments of f and g in the integral in Eq. 10.48 can be reversed. This integral is known as the "convolution" or "convolution integral" of f and g.

Let us now return to Eq. 10.4. If a function $f(t)$ is bounded near $t = 0$, then

$$\lim_{s\to\infty} F(s) = 0. \qquad (10.49)$$

If the derivative of f is also bounded near $t = 0$, then Eq. 10.4 yields the "initial value theorem"

$$\lim_{s\to\infty} \mathscr{L}\left\{\frac{df}{dt}\right\} = 0 = \lim_{s\to\infty} sF(s) - f(0+) \qquad (10.50)$$

or

$$f(0+) = \lim_{s\to\infty} sF(s). \qquad (10.51)$$

We do not propose to pursue this point; instead we investigate the fact that the behavior of $F(s)$ for large values of s is related to the behavior of $f(t)$ for small values of t. To illustrate this, we treat the transform given in Eq. 10.5 for $f(t) = e^{at}$:

$$\mathscr{L}\{e^{at}\} = F(s) = \frac{1}{s-a}. \qquad (10.52)$$

The expansion of this for large values of s is

$$F(s) = \frac{1}{s}\frac{1}{1-\dfrac{a}{s}} \to \frac{1}{s}\left(1 + \frac{a}{s} + \frac{a^2}{s^2} + \frac{a^3}{s^3} + \cdots\right). \qquad (10.53)$$

If we invert this term by the term according to Eq. 10.39, we obtain

$$f(t) = 1 + at + \frac{a^2 t^2}{2!} + \frac{a^3 t^3}{3!} + \cdots, \tag{10.54}$$

which we recognize to be the power series expansion of $f(t) = e^{at}$ about $t = 0$. This procedure of expansion of $F(s)$ for large s is useful for obtaining the behavior of $f(t)$ near $t = 0$ even when the inversion of $F(s)$ for all t is difficult or complicated.

Similarly, the behavior of $F(s)$ for small values of s is related to the asymptotic behavior of $f(t)$ for large values of t. To illustrate this, consider the transform

$$F(s) = \frac{1}{\sqrt{s}\left(1 + k\sqrt{s}\right)} = \frac{1}{\sqrt{s}} - \frac{k}{1 + k\sqrt{s}} \tag{10.55}$$

whose inverse happens to be

$$f(t) = \frac{1}{k} e^{t/k^2} \operatorname{erfc}(\sqrt{t} / k). \tag{10.56}$$

For small values of s, the first term in Eq. 10.55 is dominant and can be inverted (see Eq. 10.40) to give

$$f(t) \rightarrow 1/\sqrt{\pi t} \text{ as } t \rightarrow \infty. \tag{10.57}$$

Some care, however, is needed in the application of these asymptotic methods (see Problem 10.2 and Chapters 19 and 20).

Finally, let us reflect on the possibility of using formula 10.42 to treat some linear differential equations with variable coefficients. Generally, it is impossible to express the Laplace transform of $g(t)f(t)$ in terms of the Laplace transforms $F(s)$ and $G(s)$. However, Eq. 10.42 allows us to write down the transform of the product if $g(t)$ is of the form t^n. In this case, though we obtain a differential equation for transforms, and this may be more difficult to solve than the original problem, aside from the question of inverting the resulting transform.

Consider the differential equation

$$\frac{d^2 y}{dt^2} + 2t \frac{dy}{dt} - 2y = 0, \tag{10.58}$$

which was treated in Chapter 6. The Laplace transform of this equation is

or

$$s^2Y(s) - sy(0) - y'(0) - 2\frac{d}{ds}[sY(s) - y(0)] - 2Y(s) = 0 \quad (10.59)$$

$$s^2Y - sy(0) - y'(0) - 2s\frac{dY}{ds} - 4Y = 0. \quad (10.60)$$

This is a linear, first-order differential equation with the solution (see Chapter 2)

$$Y(s) = \frac{A}{s^2}e^{s^2/4} + y(0)\left[\frac{1}{s} + \frac{\sqrt{\pi}}{s^2}e^{s^2/4}\text{erfc}(s/2)\right] + \frac{y'(0)}{s^2}. \quad (10.61)$$

The term multiplied by A cannot correspond to any function and hence is rejected (i.e., set $A = 0$). The remaining terms are inverted to give the general solution to Eq. 10.58:

$$y(t) = y(0)\left[e^{-t^2} + \sqrt{\pi}t\,\text{erf}(t)\right] + y'(0)t. \quad (10.62)$$

Thus, we see that the Laplace transformation can be applied to some linear differential equations with variable coefficients but that the method is not of general utility and can make the problem more complicated.

Problems

10.1 Evaluate the Laplace transform of $t\delta(t-1)$ by direct calculation and by means of Eq. 10.42 with $\mathcal{L}\{\delta(t-1)\} = e^{-s}$. Discuss any difficulties or contradictions encountered.

10.2 Discuss the problems associated with inverting $\frac{1}{s-a}$ to obtain a solution valid for large t. You might recall that the transform existed only for $s > a$.

10.3 Verify that the function $f(t)$ given by Eq. 10.56 approaches the asymptotic form given by Eq. 10.57 as $t \to \infty$.

10.4 Investigate whether the Laplace transform method is of any help in solving the equations

(a) $\dfrac{d^2y}{dt^2} + 2t\dfrac{dy}{dt} + 2y = 0.$

(b) $\dfrac{d^2y}{dt^2} + 2\dfrac{dy}{dt} - 2ty = 0.$

Chapter 11

The Sturm–Liouville System

The Sturm–Liouville system comprises a differential equation of the form

$$\frac{d}{dx}\left[p(x)\frac{dy}{dx}\right]+[\lambda r(x)-q(x)]y=0, \qquad (11.1)$$

valid over the interval $[a, b]$, subject to the boundary conditions

$$\alpha y - \alpha_1 y' = 0 \quad \text{at} \quad x = a$$

$$\beta y + \beta_1 y' = 0 \quad \text{at} \quad x = b, \qquad (11.2)$$

with either α or $\alpha_1 \neq 0$ and β or $\beta_1 \neq 0$. The values of p, q, r, α, α_1, β, and β_1 are specified and real.

This system arises frequently in convective diffusion and potential theory problems, subsequent to separation of the variables. Many higher transcendental equations, such as Legendre's and Bessel's equations, are examples of the differential equation governing the Sturm–Liouville system. Our discussion here will cover some of the interesting properties of the system, which are useful in obtaining the general solution to problems. Most of the properties are presented without proof.

1. If λ is chosen arbitrarily, there is only the trivial solution $y(x) = 0$, since the equation and the boundary conditions are homogeneous. However, careful choice of λ so as to satisfy the boundary conditions yields an infinite set of characteristic or

The Newman Lectures on Mathematics
John Newman and Vincent Battaglia
Copyright © 2018 Pan Stanford Publishing Pte. Ltd.
ISBN 978-981-4774-25-3 (Hardcover), 978-1-315-10885-8 (eBook)
www.panstanford.com

eigenvalues λ_n and nontrivial solutions to Eq. 11.1, known as eigenfunctions $y_n(x)$.

2. If $r(x)$ does not change sign in $[a, b]$, then the λ_n will all be real.

3. If $p(x)$ and $r(x)$ are positive, continuous functions in $[a, b]$, which may be zero at a finite number of points, and if $q(x) \geq 0$ and is continuous, and if both $\alpha\alpha_1$ and $\beta\beta_1 \geq 0$, then the eigenvalues λ_n are all positive and increase monotonically toward infinity, and the eigenfunctions $y_n(x)$ have exactly $n - 1$ zeroes in the open interval (a, b).

4. Asymptotic solutions for eigenvalues and eigenfunctions, valid for large n, are available and often helpful [1].

5. Products of distinct eigenfunctions weighted with the function $r(x)$ may be integrated over $[a, b]$ with the result

$$\int_a^b r(x)y_n(x)y_m(x)dx = 0 \quad \text{for} \quad n \neq m. \tag{11.3}$$

Equivalently, we say that the eigenfunctions corresponding to distinct eigenvalues λ_n are *orthogonal* on $[a, b]$ with respect to the *weight function* $r(x)$.

6. An arbitrary function $f(x)$, which is well behaved on $[a, b]$, can be expanded in an infinite series of eigenfunctions:

$$f(x) = \sum_{n=1}^{\infty} a_n y_n(x), \tag{11.4}$$

where the coefficients a_n, known as Fourier expansion coefficients, can be found by using the orthogonality conditions of eigenfunctions. If we multiply Eq. 11.4 by $r(x)$ $y_m(x)$ and integrate from a to b, we have

$$\int_a^b r(x)f(x)y_m(x)dx = \sum_{n=1}^{\infty} a_n \int_a^b r(x)y_n(x)y_m(x)dx. \tag{11.5}$$

The right side of Eq. 11.5 reduces to

$$\int_a^b r(x)f(x)y_m(x)dx = a_m \int_a^b r(x)y_m^2(x)dx, \tag{11.6}$$

since by Eq. 11.3, the only nonzero term in the summation occurs when $n = m$. Thus, we have that the expansion coefficients are given by

$$a_n = \frac{\int\limits_a^b r(x)f(x)y_n(x)dx}{\int\limits_a^b r(x)y_n^2(x)dx}. \tag{11.7}$$

Reference

1. Milton Abramowltz and Irene A. Stegun (Eds.). *Handbook of Mathematical Functions with Formulas, Graphs, and Mathematical Tables*, Washington, D. C.: National Bureau of Standards, 1964, pp. 450–452.

Problems

11.1 Derive the orthogonality condition, Eq. 11.3.

11.2 Determine approximations to the first root of $J_0(\mu) = 0$ by applying the method of Stodola and Vianello to the problem

$$x^2 \frac{d^2 y}{dx^2} + x \frac{dy}{dx} + \mu^2 x^2 y = 0,$$

y is finite at $x = 0$, $\quad y = 0$ at $x = 1$.

Note: The method of Stodola and Vianello is not defined in this book. Readers would need to look it up in one of the suggested references or online.

Chapter 12

Numerical Methods for Ordinary Differential Equations

We have looked extensively at the properties of solutions to ordinary differential equations and at methods of obtaining analytic solutions. However, the state of the art today is such that most differential equations can be solved easily by standard numerical techniques. The procedures are essentially different for initial value problems and boundary value problems, that is, where the boundary conditions are all specified at one point as opposed to conditions at two points. In both cases, there are techniques that can handle several coupled differential equations, and nonlinear systems are usually amenable to analysis.

Because I am more familiar with them, I shall confine myself here to boundary value problems and use the Graetz problem as an example.

$$\frac{d^2 R}{dx^2} + \frac{1}{x}\frac{dR}{dx} + \lambda^2 (1 - x^2) R = 0 \,. \tag{12.1}$$

$dR/dx = 0$, $R = 1$ at $x = 0$, $R = 0$ at $x = 1$.

Since λ^2 is unknown, the problem is nonlinear. If λ^2 were selected at random, we could not specify these three boundary conditions since this is an eigenvalue problem. I have recently extended the method of finite differences to such problems. A point of interest here is that

The Newman Lectures on Mathematics
John Newman and Vincent Battaglia
Copyright © 2018 Pan Stanford Publishing Pte. Ltd.
ISBN 978-981-4774-25-3 (Hardcover), 978-1-315-10885-8 (eBook)
www.panstanford.com

we know in advance that the solution to the problem as stated is not unique, since there are infinitely many eigenvalues.

The way to treat nonlinear problems is to linearize them about a trial solution, λ_0^2 and $R_0(x)$, and then to iterate, or repeat the calculation, until the answers do not change. When linearized about the trial solution, Eq. 12.1 becomes

$$\frac{d^2R}{dx^2} + \frac{1}{x}\frac{dR}{dx} + \lambda_0^2 R\left(1-x^2\right) + \lambda^2 R_0\left(1-x^2\right) = \lambda_0^2 R_0\left(1-x^2\right). \quad (12.2)$$

The only nonlinear term is $\lambda^2 R$. We, of course, take λ^2 to be an unknown instead of λ. We write

$$\lambda^2 = \lambda_0^2 + \lambda_1^2 \text{ and } R = R_0 + R_1, \quad (12.3)$$

where λ_1^2 and R_1 are presumably small corrections. Thus,

$$\lambda^2 R = \lambda_0^2 R_0 + \lambda_1^2 R_0 + \lambda_0^2 R_1 + \lambda_1^2 R_1. \quad (12.4)$$

The quadratic term $\lambda_1^2 R_1$ should be very small and is discarded. Usually, I replace the corrections R_1 and λ_1^2 in favor of the original unknowns so that

$$\lambda^2 R = \lambda_0^2 R_0 + R_0\left(\lambda^2 - \lambda_0^2\right) + \lambda_0^2\left(R - R_0\right) = R_0\lambda^2 + \lambda_0^2 R - \lambda_0^2 R_0,$$

$$(12.5)$$

which is what we see in Eq. 12.2. Sometimes it is advantageous to solve for the corrections in order to avoid instabilities due to truncation error (carrying a finite number of significant figures in the calculations).

The next thing to do is to put the differential equation into finite difference form. From $x = 0$ to $x = 1$, we establish NJ "mesh points" spaced with a mesh interval h:

$$h = 1/(NJ - 1). \quad (12.6)$$

These are the points at which we carry out the calculations. In order to approximate the differential equation to order h^2, we use central difference approximations:

$$\frac{dR}{dx}\bigg|_{x=x_j} = \frac{R_{j+1} - R_{j-1}}{2h} + O(h^2); \quad \frac{d^2R}{dx^2}\bigg|_{x=x_j} = \frac{R_{j+1} + R_{j-1} - 2R_j}{h^2} + O(h^2).$$

$$(12.7)$$

Equation 12.2 thus becomes

$$\frac{R_{j+1}+R_{j-1}-2R_j}{h^2}+\frac{1}{x_j}\frac{R_{j+1}-R_{j-1}}{2h}$$

$$+\lambda_0^2 R_j\left(1-x_j^2\right)+\lambda^2 R_{0j}\left(1-x_j^2\right)=\lambda_0^2 R_{0j}\left(1-x_j^2\right). \qquad (12.8)$$

If we multiply through by h^2 and collect terms, we obtain

$$R_{j-1}\left[1-\frac{h}{2x_j}\right]+R_j\left[-2+\lambda_0^2 h^2\left(1-x_j^2\right)\right]+R_{j+1}\left[1+\frac{h}{2x_j}\right]$$

$$+\lambda^2\left[R_{0j}h^2\left(1-x_j^2\right)\right]=\lambda_0^2 R_{0j}h^2\left(1-x_j^2\right). \qquad (12.9)$$

This equation applies to points $j = 2$ to $NJ - 1$. At $j = NJ$, we write

$$R_j = 0, j = NJ \qquad (12.10)$$

to express the boundary condition.

At the center of the pipe, we first write

$$R_j = 1, j = 1. \qquad (12.11)$$

Instead of trying to program the boundary condition $dR/dx = 0$ at $x = 0$, we recognize that $x = 0$ is a singular point of the differential equation, in fact, the equation behaves like Bessel's equation of order zero near $x = 0$. Consequently, since we want the regular solution, we expand R in a power series about $x = 0$:

$$R = 1+\frac{1}{2}R''(0)x^2 +0\left(x^4\right). \qquad (12.12)$$

(As you can readily verify, R is even in x, and the third derivative is also zero at $x = 0$.) Substitution into the differential equation gives

$$R''(0)+\frac{1}{x}R''(0)x+\lambda^2 +0\left(x^2\right)=0. \qquad (12.13)$$

Note that while R' is zero at $x = 0$, the term $\frac{1}{x}R'$ still makes an appreciable contribution at $x = 0$. This is related to the fact that $x = 0$ is a singular point of the equation. Equation 12.13 gives $R''(0)=-\frac{1}{2}\lambda^2$, and Eq. 12.12 then yields the correct finite difference expression of the boundary condition $R'(0) = 0$:

$$R_{j+1} + \frac{1}{4}h^2\lambda^2 = 1 + 0\left(h^4\right) \text{ at } j = 1.$$ (12.14)

Because I wanted to use standard techniques for coupled ordinary differential equations, I next added the differential equation

$$d\lambda^2/dx = 0,$$ (12.15)

with the finite difference form

$$\lambda_j^2 = \lambda_{j-1}^2, \quad j = 2, \ldots, NJ.$$ (12.16)

There is no boundary condition for Eq. 12.15, but this is what allows us to write the three boundary conditions for Eq. 12.1.

We now solve the difference equations. I shall not go into the method here; it is written in the August 1968 issue of *Industrial and Engineering Chemistry Fundamentals* and in Appendix C of *Electrochemical Systems* (Wiley, 2004). We note that we have $2NJ$ linear equations for the $2NJ$ unknowns, λ_j^2 and R_j at the mesh points. It is not necessary to use general matrix methods, however, since each equation involves at most three mesh points. This method is described in the above mentioned sources.

After we obtain λ^2 and R, we use these values for λ_0^2 and R_0 and repeat the calculation until it converges, which invariably takes less than 10 iterations if it works at all. The convergence is indicated below by showing λ^2 for successive iterations, in comparison with the method of Stodola and Vianello.

Linearized problem	Stodola and Vianello
7.113766	5.33
7.312204	6.8844
7.313463	7.2329
7.313463	

The convergence is very rapid for a problem that has been properly linearized and is said to be quadratic in the sense that if the error is ε at one step, it is proportional to ε^2 at the next. This has the property of doubling the number of correct significant figures with each step. By contrast, a method with a linear convergence has an error proportional to ε after the next step.

We mentioned before that this problem does not have a unique solution. Consequently, we took as a first approximation

$$R_0 = \cos\left[\left(\ell + \frac{1}{2}\right)\pi x\right], \quad \ell = 0, 1, 2, 3, 4, \ldots$$ (12.17)

which has the right number of oscillations and which satisfies the boundary conditions. The initial guess for λ^2 we took from

$$\lambda_0^2 = -2R_0''(0) = 2\left(\ell + \frac{1}{2}\right)^2 \pi^2. \tag{12.18}$$

(A better guess might have been $\lambda_0^2 = (4\ell + 8/3)^2$, which we know from the asymptotic solution of Sellars, Tribus, and Klein [1].) With these trial solutions, the problem converged readily for the first five eigenvalues and eigenfunctions, all that were attempted.

In a numerical problem of this type, inaccuracies can arise from three sources:

1. Convergence of the nonlinear problem. With proper linearization and a quadratic convergence, this is no problem.
2. Truncation errors from carrying a finite number of significant figures. With the old IBM machines, which fortunately we have replaced, this was sometimes a problem and for precise work, it might be necessary to use "double precision" calculations.
3. Errors arising from the finite difference approximation to the differential equations. These generally become smaller as h is decreased. It is a test of our programming to see whether our answers are accurate to the order h^2, as we claimed. This is indicated below.

h	λ^2	Δ	Extrapolation to $h = 0$
0.01	7.31309262		
		0.00037072	7.313586913
0.005	7.31346334		
		0.00009269	7.313586927
0.0025	7.31355603		
			7.3135868
			(Abramowitz)

Even if one had unlimited storage space and computation time, errors from truncation and from the finite difference approximation work against each other so that the error will be a minimum for an optimum value of h, which is neither large nor small.

The solution to the Graetz problem as outlined here took about 4 h of time, including programming, keypunching, feeding the computer, and debugging, but not including the contemplation and refinement of the results. The eigenvalues obtained are as good as those in the literature; the functions and coefficients for calculating the mass transfer rates are considerably better. The solutions for the first three eigenfunctions and eigenvalues are given in Fig. 12.1.

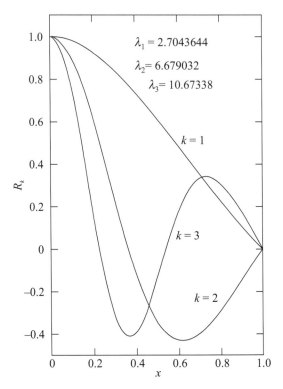

$\lambda_1 = 2.7043644$

$\lambda_2 = 6.679032$

$\lambda_3 = 10.67338$

$k = 1$

$k = 3$

$k = 2$

Figure 12.1 Graetz functions.

Reference

1. John Randolph Sellars, Myron Tribus, and John Klein. "Heat transfer to laminar flow in a round tube or flat conduit: the Graetz problem extended." *Transactions of the American Society of Mechanical Engineers*, **78**, 441–447 (1956).

Chapter 13

Vector Calculus

We find it convenient to use vectors for the simple reason that many physical quantities are vectors, for example, velocity, force, and acceleration. It is assumed here that you have some prior knowledge of vectors.

Newton's second law of motion can be written in the vector notation as

$$\underline{F} = m\underline{a} \tag{13.1}$$

instead of writing out the component form

$$F_x = ma_x, F_y = ma_y, F_z = ma_z. \tag{13.2}$$

It is a consequence of the fact that force and acceleration are vectors that this law can be written more compactly in vector notation. Furthermore, Eq. 13.1 is valid without reference to any particular coordinate system, whereas Eq. 13.2 can lead to such irrelevant questions as "what is x?" or "in what direction is x?"

Let us consider the equation describing conservation of a quantity. For a conserved quantity, the rate of accumulation in any given volume is equal to the net rate of input of that quantity, as you learned in your elementary courses on energy and material balances. The net rate of input is determined by the flux of that quantity, and the flux is a vector. Let the quantity per unit volume be denoted by c,

The Newman Lectures on Mathematics
John Newman and Vincent Battaglia
Copyright © 2018 Pan Stanford Publishing Pte. Ltd.
ISBN 978-981-4774-25-3 (Hardcover), 978-1-315-10885-8 (eBook)
www.panstanford.com

and the flux or rate of flow of that quantity per unit area and per unit time, by $\underline{N}$. Thus, if the quantity is mass,

$$c = \rho, \text{ the density} \tag{13.3}$$

and

$$\underline{N} = \rho\underline{v}, \tag{13.4}$$

where $\underline{v}$ is the mass average velocity.

Consider a volume element $\Delta x \Delta y \Delta z$, as sketched in Fig. 13.1. Then conservation of the quantity is expressed as

$$\frac{d}{dt}[c\Delta x \Delta y \Delta z] = \left(N_x\big|_x - N_x\big|_{x+\Delta x}\right)\Delta y \Delta z + \left(N_y\big|_y - N_y\big|_{y+\Delta y}\right)\Delta x \Delta z$$

$$+ \left(N_z\big|_z - N_z\big|_{z+\Delta z}\right)\Delta x \Delta y. \tag{13.5}$$

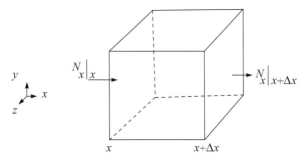

Figure 13.1 Volume element for assessing material balances.

Divide throughout by $\Delta x \Delta y \Delta z$ and take the limit as Δx, Δy, and Δz approach zero.

$$\frac{\partial c}{\partial t} = \lim_{\Delta x \to 0} \frac{N_x\big|_x - N_x\big|_{x+\Delta x}}{\Delta x} + \lim_{\Delta y \to 0} \frac{N_y\big|_y - N_y\big|_{y+\Delta y}}{\Delta y}$$

$$+ \lim_{\Delta z \to 0} \frac{N_z\big|_z - N_z\big|_{z+\Delta x}}{\Delta z}. \tag{13.6}$$

We recall the definition of a derivative, $\dfrac{\partial p}{\partial x} = \lim\limits_{\Delta x \to 0} \dfrac{\left(p\big|_{x+\Delta x} - p\big|_x\right)}{\Delta x}$, and therefore are able to write Eq. 13.6 as

$$\frac{\partial c}{\partial t} = -\frac{\partial N_x}{\partial x} - \frac{\partial N_y}{\partial y} - \frac{\partial N_z}{\partial z}. \tag{13.7}$$

This equation then expresses the conservation of the quantity. Since the conservation law should be independent of the choice of the coordinate system, we are led to define the following vector operation:

$$\nabla \cdot \underline{N} = \frac{\partial N_x}{\partial x} + \frac{\partial N_y}{\partial y} + \frac{\partial N_z}{\partial z}. \tag{13.8}$$

The quantity $\nabla \cdot \underline{N}$ is called the *divergence* of the vector $\underline{N}$ and is seen to be a scalar. The physical significance of the divergence is best seen from the conservation equation itself, which becomes

$$\partial c / \partial t = -\nabla \cdot \underline{N}. \tag{13.9}$$

The quantity $-\nabla \cdot \underline{N}$ represents the net flux into a differential volume element and might logically be called the "convergence." We also see, as might be expected, that the conservation law becomes more compact when written in the form 13.9 and, by design, contains no extraneous reference to any coordinate system.

Now let us regard c as the concentration of a species and let the flux express the fact that this species diffuses from regions of higher concentration to regions of lower concentration:

$$N_x = -\mathcal{D}\frac{\partial c}{\partial x}, \ N_y = -\mathcal{D}\frac{\partial c}{dy}, \ N_z = -\mathcal{D}\frac{\partial c}{\partial z}, \tag{13.10}$$

where $\mathcal{D}$ is known as the diffusion coefficient. We write this diffusion law more compactly and independent of the coordinate system as

$$\underline{N} = -\mathcal{D}\nabla c, \tag{13.11}$$

where

$$\nabla c = \underline{e}_x \frac{\partial c}{\partial x} + \underline{e}_y \frac{\partial c}{\partial y} + \underline{e}_z \frac{\partial c}{\partial z}, \tag{13.12}$$

where $\underline{e}_x$, $\underline{e}_y$, and $\underline{e}_z$ are unit (dimensionless) vectors in the x-, y-, and z-directions, respectively. ∇c is a vector quantity and is known as the *gradient* of the scalar c.

If $\mathcal{D}$ is a constant, then substitution of Eq. 13.10 in Eq. 13.7 yields

$$\frac{\partial c}{\partial t} = \mathcal{D}\left(\frac{\partial^2 c}{\partial x^2} + \frac{\partial^2 c}{\partial y^2} + \frac{\partial^2 c}{\partial z^2} \right). \tag{13.13}$$

This gives us a partial differential equation for c from which, with appropriately stated boundary conditions, we might hope to determine c. It also leads us to define the Laplacian, a scalar, of the scalar c

$$\nabla^2 c = \nabla \cdot \nabla c = \frac{\partial^2 c}{\partial x^2} + \frac{\partial^2 c}{\partial y^2} + \frac{\partial^2 c}{\partial z^2}, \qquad (13.14)$$

so that Eq. 13.13 can be written as

$$\frac{\partial c}{\partial t} = \mathcal{D}\nabla^2 c. \qquad (13.15)$$

We should also define the *curl* of a vector field:

$$\nabla \times \underline{v} = \begin{vmatrix} \underline{e}_x & \underline{e}_y & \underline{e}_z \\ \partial/\partial x & \partial/\partial y & \partial/\partial z \\ v_x & v_y & v_z \end{vmatrix}$$

$$= \underline{e}_x \left(\frac{\partial v_z}{\partial y} - \frac{\partial v_y}{\partial z} \right) + \underline{e}_y \left(\frac{\partial v_x}{\partial z} - \frac{\partial v_z}{\partial x} \right) + \underline{e}_z \left(\frac{\partial v_y}{\partial x} - \frac{\partial v_x}{\partial y} \right).$$

$$(13.16)$$

This quantity arises frequently in fluid mechanics and in electromagnetic field theory. Its physical significance can be seen from the fact that if $\underline{v}$ is the fluid velocity, then $\nabla \times \underline{v}$, which is called the vorticity, is equal to twice the angular velocity of rotation of a differential fluid element, say if it were isolated and allowed to equilibrate in a frictionless enclosure with no change in angular momentum. A fluid motion for which $\nabla \times \underline{v} = 0$ is consequently said to be "irrotational."

Let us return to the law of conservation of mass, which can be written as

$$\frac{\partial \rho}{\partial t} = -\nabla \cdot (\rho \underline{v}). \qquad (13.17)$$

Let us differentiate the product in component form:

$$\nabla \cdot (\rho \underline{v}) = \frac{\partial \rho v_x}{\partial x} + \frac{\partial \rho v_y}{\partial y} + \frac{\partial \rho v_z}{\partial z} = v_x \frac{\partial \rho}{\partial x} + v_y \frac{\partial \rho}{\partial y} + v_z \frac{\partial \rho}{\partial z}$$

$$+ \rho \frac{\partial v_x}{\partial x} + \rho \frac{\partial v_y}{\partial y} + \rho \frac{\partial v_z}{\partial z}.$$

$$(13.18)$$

We recognize that the first three terms on the right represent the dot product of the vector $\underline{v}$ and the vector $\nabla\rho$. (We expect you to have prior knowledge of dot products.) Consequently, we have derived the important identity for differentiation of products of a scalar and a vector:

$$\nabla\cdot(\rho\underline{v}) = \underline{v}\cdot\nabla\rho + \rho\nabla\cdot\underline{v} \qquad (13.19)$$

We notice that we started with a scalar $\nabla\cdot(\rho\underline{v})$ and obtained two quantities that are scalars. The appendix to this chapter gives a number of vector identities that make it easier to manipulate vector equations without resort to derivations in component form as we have used here.

The law of conservation of mass can thus be written as

$$\frac{\partial\rho}{\partial t} = -\underline{v}\cdot\nabla\rho - \rho\nabla\cdot\underline{v} . \qquad (13.20)$$

We observe that if the density of the medium does not vary in space or time or if such variations can be neglected, then the velocity of that medium satisfies the equation

$$\nabla\cdot\underline{v} = 0, \qquad (13.21)$$

that is, the divergence of the velocity vanishes. Such a medium is said to be incompressible.

If in the expression 13.11 of the flux of a species the diffusion coefficient is not constant, then the conservation equation becomes

$$\frac{\partial c}{\partial t} = \nabla\cdot(\mathcal{D}\nabla c) = \mathcal{D}\nabla^2 c + (\nabla c)\cdot\nabla\mathcal{D} . \qquad (13.22)$$

If the diffusion coefficient depends only on the concentration c, then

$$\frac{\partial c}{\partial t} = \mathcal{D}\nabla^2 c + (\nabla c)^2\frac{d\mathcal{D}}{dc} . \qquad (13.23)$$

For diffusion in a flowing fluid, we should expect the flux expression to be modified to read

$$\underline{N} = c\underline{v} - \mathcal{D}\nabla c, \qquad (13.24)$$

that is, the species is convected in addition to diffusing. The conservation law becomes

$$\frac{\partial c}{\partial t} + \underline{v}\cdot\nabla c + c\nabla\cdot\underline{v} = \nabla\cdot(\mathcal{D}\nabla c) . \qquad (13.25)$$

For an incompressible fluid with a constant diffusion coefficient, we thus have the equation of convective diffusion

$$\frac{\partial c}{\partial t} + \underline{v} \cdot \nabla c = \mathcal{D} \nabla^2 c. \tag{13.26}$$

If the species is not conserved, but is produced at a rate R per unit volume and per unit time, then the conservation equation must be modified to read

$$\frac{\partial c}{\partial t} = -\nabla \cdot \underline{N} + R. \tag{13.27}$$

This equation says that the rate of accumulation per unit volume is equal to the net rate of input plus the volumetric rate of production. R would be expected to be expressed by the appropriate laws of chemical kinetics. For example, if the species reacts (that is, *disappears*) according to a first-order rate law, then

$$R = -kc. \tag{13.28}$$

Let us now express the law of conservation of mass for a volume element of some extent:

$$\frac{d}{dt} \int \rho dV = -\oint \rho \underline{v} \cdot \underline{dS} \tag{13.29}$$

where $\underline{dS}$ represents a surface element vector having the direction of the outward normal to the surface and the surface integral is over the closed surface of the volume element. The net rate of flow of mass out of the volume element is thus obtained by taking the dot product of the mass flux $\rho\underline{v}$ with the outward normal unit vector and integrating over the surface of the volume element. On the other hand, the integration of the differential conservation law 13.17 over the volume of the element yields

$$\frac{d}{dt} \int \rho dV = -\int \nabla \cdot (\rho\underline{v}) dV. \tag{13.30}$$

Consequently, we infer the important *divergence theorem*

$$\oint \underline{N} \cdot \underline{dS} = \int \nabla \cdot \underline{N} dV \tag{13.31}$$

for an arbitrary volume element. The rigorous derivation contains no concepts not already expressed here in a physical context.

At the same time, we may mention, but not derive, Stokes's theorem

$$\oint \underline{v} \cdot d\underline{\ell} = \int (\nabla \times \underline{v}) \cdot d\underline{S} \qquad (13.32)$$

for a line integral along a closed curve with an appropriate convention for the directions of $d\underline{\ell}$ and $d\underline{S}$, and Green's theorem

$$\int \left(\psi \nabla^2 \phi - \phi \nabla^2 \psi \right) dV = \oint \left(\psi \nabla \phi - \phi \nabla \psi \right) \cdot d\underline{S}, \qquad (13.33)$$

which can be derived from the divergence theorem.

Let us evaluate the curl of the gradient of a scalar ϕ from Eqs. 13.12 and 13.16:

$$\nabla \times \nabla \phi = \underline{e}_x \left(\frac{\partial}{\partial y} \frac{\partial \phi}{\partial z} - \frac{\partial}{\partial z} \frac{\partial \phi}{\partial y} \right) + \underline{e}_y \left(\frac{\partial}{\partial z} \frac{\partial \phi}{\partial x} - \frac{\partial}{\partial x} \frac{\partial \phi}{\partial z} \right)$$
$$+ \underline{e}_z \left(\frac{\partial}{\partial x} \frac{\partial \phi}{\partial y} - \frac{\partial}{\partial y} \frac{\partial \phi}{\partial x} \right). \qquad (13.34)$$

Since the order of differentiation can be interchanged, it follows that the curl of the gradient of any scalar is identically zero. The converse is also true; any vector whose curl is identically zero can always be expressed as the gradient of a scalar, which is uniquely determined except for an additive constant. Consequently, for an irrotational flow of a fluid

$$\nabla \times \underline{v} = 0, \qquad (13.35)$$

the velocity can always be expressed as the gradient of a scalar "velocity potential"

$$\underline{v} = \nabla \phi. \qquad (13.36)$$

For the irrotational flow of a fluid that happens to be incompressible, we have the interesting result that the flow field can be determined by solving Laplace's equation for the velocity potential,

$$\nabla^2 \phi = 0, \qquad (13.37)$$

subject, of course, to appropriate boundary conditions.

Also, in electrostatics, the curl of the electric field is zero:

$$\nabla \times \underline{E} = 0, \qquad (13.38)$$

which means that the electric field can be expressed in terms of the gradient of an electrostatic potential

$$\underline{E} = -\nabla \Phi. \qquad (13.39)$$

Furthermore, the electric field also satisfies the equation

$$\nabla \cdot (\varepsilon \underline{E}) = \rho_e. \tag{13.40}$$

where ρ_e is the electric charge density and ε is the permittivity of the medium. Thus, for a medium of uniform permittivity, the electrostatic potential satisfies Poisson's equation

$$\nabla^2 \Phi = -\rho_e / \varepsilon, \tag{13.41}$$

and if the medium is also charge free, the electrostatic potential satisfies Laplace's equation

$$\nabla^2 \Phi = 0. \tag{13.42}$$

If we set $\underline{v} = \nabla \phi$, then we can combine Eqs. 13.32 and 13.34 to obtain

$$\oint (\nabla \phi) \cdot \underline{d\ell} = \int (\nabla \times \nabla \phi) \cdot \underline{dS} = 0. \tag{13.43}$$

This tells us that the integral of the gradient around a closed curve yields zero, which we should certainly hope to be true if ϕ is to be single valued.

Finally, we should mention that quantities exist, which are neither scalars nor vectors but are called tensors. For example, the equation of conservation of momentum (the differential expression of Newton's second law of motion) is

$$\rho \left(\frac{\partial \underline{v}}{\partial t} + \underline{v} \cdot \nabla \underline{v} \right) = -\nabla p - \nabla \cdot \underline{\underline{\tau}} + \rho \underline{g}. \tag{13.44}$$

This is a vector equation involving the density ρ, velocity $\underline{v}$, pressure p, stress $\underline{\underline{\tau}}$, and gravitational acceleration $\underline{g}$. Note that the gradient $\nabla \underline{v}$ of a vector $\underline{v}$ is not a vector since it is necessary to specify how each of the components v_x, v_y, v_z varies in each of the directions x, y, and z:

$$(\nabla \underline{v})_{xx} = \frac{\partial v_x}{\partial x}, \quad (\nabla \underline{v})_{xy} = \frac{\partial v_y}{\partial x}, \quad \text{etc.} \tag{13.45}$$

Also the stress $\underline{\underline{\tau}}$ is a tensor whose nine components tell us the force (a vector) acting on surfaces with various orientations (specified by a normal unit vector). The stress is symmetric (while the gradient of the velocity need not be) and, for a Newtonian fluid, is given for the diagonal elements by

$$\tau_{xx} = -2\mu\frac{\partial v_x}{\partial x} + \frac{2}{3}\mu\nabla\cdot\underline{v}, \text{ etc.} \qquad (13.46)$$

and for the off-diagonal elements by

$$\tau_{xy} = -\mu\left(\frac{\partial v_x}{\partial y} + \frac{\partial v_y}{\partial x}\right), \text{ etc.} \qquad (13.47)$$

where μ is the viscosity. We have not defined here any operations involving tensors; they are given in the appendix.

Appendix: Vector and Tensor Algebra and Calculus

1. Definitions
    (a) Dyadic product: $(\underline{a}\,\underline{c})_{ij} = a_i c_j$ ($\underline{a}\,\underline{c}$ is a tensor.)
    (b) Double dot product:
    $$\underline{\underline{\sigma}} : \underline{\underline{\tau}} = \sum_i\sum_j \sigma_{ij}\tau_{ji}$$

    (c) A tensor operating on a vector from the right yields a vector:
    $$\underline{a}\cdot\underline{\underline{\tau}} = \sum_i\sum_j \underline{e}_i a_j \tau_{ji}$$

    (d) Transpose of a tensor: $\left(\underline{\underline{\tau}}^*\right)_{ij} = \tau_{ji}$ or $\underline{\underline{\tau}}\cdot\underline{a} = \underline{a}\cdot\underline{\underline{\tau}}^*$

    (e) Product of two tensors:
    $$\left(\underline{\underline{\tau}}\cdot\underline{\underline{\sigma}}\right)\cdot\underline{v} = \underline{\underline{\tau}}\cdot\left(\underline{\underline{\sigma}}\cdot\underline{v}\right) \text{ or}$$
    $$\left(\underline{\underline{\tau}}\cdot\underline{\underline{\sigma}}\right)_{ij} = \sum_k \tau_{ik}\sigma_{kj}$$

    (f) The divergence of a tensor is a vector: $\nabla\cdot\underline{\underline{\tau}} = \sum_i\sum_j \underline{e}_i \frac{\partial\tau_{ji}}{\partial x_j}$

    (g) Laplacian of a scalar: $\nabla^2\Phi = \nabla\cdot\nabla\Phi = \sum_i \frac{\partial^2\Phi}{\partial x_i^2}$

    (h) Gradient of a vector: $(\nabla\underline{v})_{ij} = \partial v_j/\partial x_i$
    (i) Laplacian of a vector:
    $\nabla^2\underline{v} = \nabla\cdot\nabla\underline{v} = \nabla(\nabla\cdot\underline{v}) - \nabla\times\nabla\times\underline{v}$

2. Algebra

(a) $\underline{\underline{\tau}}:(\underline{a}\,\underline{b})=\underline{b}\cdot\left(\underline{\underline{\tau}}\cdot\underline{a}\right)$

(b) $(\underline{u}\,\underline{v}):(\underline{w}\,\underline{z})=(\underline{u}\,\underline{w}):(\underline{v}\,\underline{z})=(\underline{u}\cdot\underline{z})(\underline{v}\cdot\underline{w})$

(c) $\underline{a}\cdot(\underline{b}\,\underline{c})=(\underline{a}\cdot\underline{b})\underline{c}$

(d) $(\underline{a}\,\underline{b})\cdot\underline{c}=\underline{a}(\underline{b}\cdot\underline{c})$

(e) $\underline{a}\times(\underline{b}\times\underline{c})=\underline{b}(\underline{a}\cdot\underline{c})-\underline{c}(\underline{a}\cdot\underline{b})$

(f) $\underline{v}\cdot(\underline{v}\times\underline{w})=\underline{v}\cdot(\underline{w}\times\underline{u})$

(g) $(\underline{u}\times\underline{v})\cdot(\underline{w}\times\underline{z})=(\underline{u}\cdot\underline{w})(\underline{v}\cdot\underline{z})-(\underline{u}\cdot\underline{z})(\underline{v}\cdot\underline{w})$

(h) $\underline{v}\cdot\left(\underline{\underline{\tau}}^{*}\cdot\underline{w}\right)=\underline{w}\cdot\left(\underline{\underline{\tau}}\cdot\underline{v}\right)$

3. Differentiation of products

(a) $\nabla\phi\psi=\phi\nabla\psi+\psi\nabla\phi$ (a vector)

(b) $\nabla\phi\underline{v}=\phi\nabla\underline{v}+(\nabla\phi)\underline{v}$ (a tensor)

(c) $\nabla(\underline{a}\cdot\underline{c})=\underline{a}\cdot\nabla\underline{c}+\underline{c}\cdot\nabla\underline{a}+\underline{a}\times\nabla\times\underline{c}+\underline{c}\times\nabla\times\underline{a}$
$\qquad\qquad=(\nabla\underline{c})\cdot\underline{a}+(\nabla\underline{a})\cdot\underline{c}$ (a vector)

(d) $\nabla\cdot(\phi\underline{v})=\phi\nabla\cdot\underline{v}+\underline{v}\cdot\nabla\phi$ (a scalar)

(e) $\nabla\cdot(\underline{v}\times\underline{w})=\underline{w}\cdot(\nabla\times\underline{v})-\underline{v}\cdot(\nabla\times\underline{w})$ (a scalar)

(f) $\nabla\times(\phi\underline{v})=\phi\nabla\times\underline{v}+(\nabla\phi)\times\underline{v}$ (a vector)

(g) $\nabla\times(\underline{b}\times\underline{c})=\underline{b}(\nabla\cdot\underline{c})-\underline{c}(\nabla\cdot\underline{b})+\underline{c}\cdot\nabla\underline{b}-\underline{b}\cdot\nabla\underline{c}$ (a vector)

(h) $\nabla\cdot(\underline{a}\underline{b})=(\nabla\cdot\underline{a})\underline{b}+\underline{a}\cdot\nabla\underline{b}$ (a vector)

(i) $\nabla\cdot(\phi\underline{\underline{\tau}})=\phi\nabla\cdot\underline{\underline{\tau}}+(\nabla\phi)\cdot\underline{\underline{\tau}}$ (a vector)

(j) $\nabla\cdot(\underline{u}\cdot\underline{\underline{\tau}})=\underline{\underline{\tau}}:\nabla\underline{u}+\underline{u}\cdot\nabla\cdot\underline{\underline{\tau}}^{*}$ (a scalar)

4. Various forms of Gauss' law (divergence theorem) and Stokes' law. (dS = area element, $d\ell$ = line element, dv = volume element)

(a) $\oint d\underline{S}\cdot\underline{F}=\int dv\nabla\cdot F$

(b) $\oint d\underline{S}\phi=\int dv\nabla\phi$

(c) $\oint(d\underline{S}\cdot\underline{G})\underline{F}=\int dv\underline{F}\nabla\cdot\underline{G}+\int dv\underline{G}\cdot\nabla\underline{F}$

(d) $\oint d\underline{S}\times\underline{F}=\int dv\nabla\times\underline{F}$

(e) $\oint d\underline{S}\cdot\underline{\underline{\tau}}=\int dv\nabla\cdot\underline{\underline{\tau}}$

(f) $\oint d\underline{S}\cdot(\psi\nabla\phi-\phi\nabla\psi)=\int dv\left(\psi\nabla^{2}\phi-\phi\nabla^{2}\psi\right)$

(g) $\oint \underline{d\ell} \cdot \underline{F} = \int \underline{dS} \cdot \nabla \times \underline{F}$

(h) $\oint \underline{d\ell}\phi = \int \underline{dS} \times \nabla \phi \, 0$

Miscellaneous

(a) $\nabla \cdot \nabla \times \underline{E} = 0$

(b) $\nabla \times \nabla \phi = 0$

(c) $\underline{w} \cdot \nabla \underline{v} = \sum_i \sum_j \underline{e}_i w_j \partial v_i / \partial x_j$

(d) $D / Dt = \partial / \partial t + \underline{v} \cdot \nabla$

(e) $\dfrac{D\underline{v}}{Dt} = \dfrac{\partial \underline{v}}{\partial t} + \nabla \left(\dfrac{1}{2} v^2 \right) - \underline{v} \times \nabla \times \underline{v}$ where $\underline{v}$ is the mass-average velocity.

Problem

13.1 Show that the irrotational flow of an incompressible Newtonian fluid identically satisfies Eq. 13.44 with $\underline{\underline{\tau}}$ given by Eqs. 13.46 and 13.47 if the curl of the gravitational acceleration is zero and if the viscosity is uniform. Indicate how to calculate the pressure distribution in such a flow field and verify that it can be calculated (except for an additive constant).

Chapter 14

Classification and Examples of Partial Differential Equations

We have studied extensively the properties and solutions to ordinary differential equations. Physical problems frequently give rise to partial differential equations; the chapter on vector calculus provides a wealth of such equations and indicated their physical origin. The solution to partial differential equations will make use of our earlier study of ordinary differential equations.

Before taking up various techniques for the solution to partial differential equations, we should consider their classification. An example of a second-order, partial differential equation is

$$a\frac{\partial^2 T}{\partial x^2} + b\frac{\partial^2 T}{\partial x \partial y} + c\frac{\partial^2 T}{\partial y^2} = 0. \tag{14.1}$$

This equation is said to be hyperbolic if $b^2 > 4ac$, parabolic if $b^2 = 4ac$, and elliptic if $b^2 < 4ac$.

This equation will be linear if a, b, and c depend only on the independent variables x and y but not on the dependent variable T. If a, b, and c are not constant, it is possible for the value of $b^2 - 4ac$ to change sign. Then Eq. 14.1 will be hyperbolic in some regions of x and y, elliptic in others, and possibly parabolic in others or on the borders between elliptic and hyperbolic regions.

A second-order, partial differential equation is said to be quasi-linear if it can be written in the form

The Newman Lectures on Mathematics
John Newman and Vincent Battaglia
Copyright © 2018 Pan Stanford Publishing Pte. Ltd.
ISBN 978-981-4774-25-3 (Hardcover), 978-1-315-10885-8 (eBook)
www.panstanford.com

$$a\frac{\partial^2 T}{\partial x^2} + b\frac{\partial^2 T}{\partial x \partial y} + c\frac{\partial^2 T}{\partial y^2} + d = 0, \tag{14.2}$$

where a, b, c, and d can depend on x, y, and T and on the first partial derivatives $\partial T/\partial x$ and $\partial T/\partial y$ but do not depend on the second partial derivatives. The classification of the equation as hyperbolic, parabolic, or elliptic still depends on the coefficients a, b, and c of the second-order derivatives, as indicated above. Since a, b, and c can depend on T, one ordinarily will not know in advance in which regions the equation is hyperbolic, parabolic, or elliptic.

If a, b, and c are constant (and $d = 0$), we can get a solution to Eq. 14.1 in the form

$$T = f(y + mx), \tag{14.3}$$

where m is a constant. Substitution into Eq. 14.1 gives

$$am^2 + bm + c = 0, \tag{14.4}$$

which usually gives two roots

$$m_1, m_2 = \frac{-b \pm \sqrt{b^2 - 4ac}}{2a}. \tag{14.5}$$

The general solution to Eq. 14.1 is then

$$T = f(y + m_1 x) + g(y + m_2 x), \tag{14.6}$$

where f and g are arbitrary functions of their arguments, arbitrary as long as they can be differentiated twice. If the two roots coincide, $m_1 = m_2$, as they would for a parabolic equation, then the general solution to Eq. 14.1 is

$$T = f(y + m_1 x) + xg(y + m_1 x). \tag{14.7}$$

The situation here is similar to that encountered in Chapter 7 when the solution to an ordinary differential equation with constant coefficients had two exponents that coincided. Note also that the arbitrary (integration) constants there become the arbitrary functions f and g here.

The solution 14.3 says that T is constant along a line in the x–y plane where $y + mx$ is constant. This is a line of slope $-m$ (see Fig. 14.1).

This classification of partial differential equations is important even if one intends to obtain a numerical solution, because the boundary conditions take different forms for the different types.

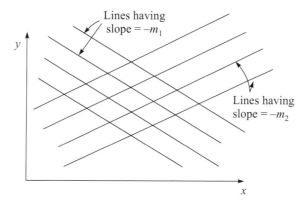

Figure 14.1 Characteristic lines in the *x–y* plane; only a hyperbolic equation will have two real values of m_1 and m_2 (if *a*, *b*, and *c* are real).

14.1 Elliptic Equations

The classic elliptic differential equation is Laplace's equation. For steady heat conduction in solids,

$$\frac{\partial^2 T}{\partial x^2} + \frac{\partial^2 T}{\partial y^2} = 0 . \tag{14.8}$$

This can be obtained from Eq. 13.15 by considering temperature to be the conserved entity, by having no time variations, and by having only two, rectangular spatial coordinates. Actually internal energy is the conserved quantity; its variation is given by $\rho \hat{C}_v dT$, where $\rho \hat{C}_v$ is the heat capacity per unit volume. An inherent assumption is that the thermal conductivity *k* is constant.

Laplace's equation could apply equally well to steady diffusion in solids. In both the heat- and mass-transfer situations, there are additional assumptions that would be apparent in a course on transport phenomena. Laplace's equation can also apply to problems of the irrotational flow of incompressible fluids. See Eq. 13.37. There is also a wealth of examples in electrochemical systems, where the electric potential satisfies Laplace's equation under certain approximations, and the solutions can describe current and potential distributions in electrodeposition and in cathodic protection of structures from corrosion.

For Laplace's equation, the slopes m_1 and m_2 are imaginary, and while Eq. 14.6 is valid, it is hardly ever applied. The characteristic lines cannot really be drawn as in Fig. 14.1. Instead, it is important for Laplace's equation, and for elliptic problems in general, that a boundary condition needs to be stated at all parts of the boundary of the (two-dimensional) region where the solution is desired. This usually takes the form of

1. A statement of the value of T along the boundary (Dirichlet condition).
2. A statement of the value of the normal derivative of T along the boundary (Neumann condition).
3. A relationship between the value of T and its normal derivative along the boundary (mixed condition).

One should think in physical terms when formulating boundary conditions. For example, consider steady heat conduction in the wall of a furnace, as sketched in Fig. 14.2. The two surfaces labeled $\partial T/\partial n = 0$ represent symmetry planes in the middle of the walls, where the normal component of the heat flux is zero. If the furnace cross section is a square, the 45° line (shown dashed) can also be a symmetry plane with zero heat flux. Then only half of the region shown in the sketch needs to be treated.

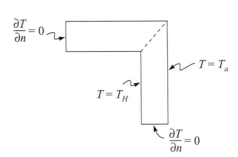

Figure 14.2 Boundary conditions for heat conduction though a furnace wall.

A simple boundary condition is shown at the other surfaces, $T = T_a$ at the outside and $T = T_H$ at the inside. If we take into account a heat-transfer coefficient h on the outside, we use a mixed condition

$$-k\frac{\partial T}{\partial n} = h(T - T_a),$$

(14.9)

where *n* is directed toward the outside. If we included radiant heat transfer on a surface, the boundary condition would become a nonlinear relationship between T and $\partial T/\partial n$.

Examples with elliptic equations are included in Chapters 15, 17, and 18.

14.2 Parabolic Equations

The classic parabolic equation is the heat equation

$$\frac{\partial T}{\partial t} = \alpha \frac{\partial^2 T}{\partial x^2} \qquad (14.10)$$

obtained from Eq. 13.15 by including the time derivative but considering only one spatial direction.

For Eq. 14.10, it is appropriate to state the initial value of T and to specify boundary conditions at two values of x. These boundary conditions can set the value of T, the value of the derivative with respect to x, or a combination of the two, much as for Laplace's equation.

For the parabolic equation, one can think that the future cannot affect the past; information propagates in one direction in time. Some flow problems may turn out to be parabolic; information may then propagate in the downstream direction. Chapter 19 deals with a parabolic problem.

14.3 Hyperbolic Equations

The characteristic lines are most useful here because two real values are obtained for *m* in Eq. 14.5.

As an example, let us develop the telegraph equation. A long transmission line has a capacitance C per unit length and an inductance L per unit length, as sketched in Fig. 14.3. The current I along the line changes because of the charging of the capacitance; for continually distributed capacitance and inductance, this takes the differential form

$$\frac{\partial I}{\partial x} = -C \frac{\partial \Phi}{\partial t} . \qquad (14.11)$$

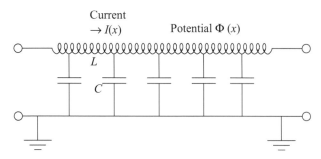

Figure 14.3 Transmission line characterized by continuously distributed inductance L and capacitance C (per unit length).

Correspondingly, the potential Φ changes with position because of a change in current with time,

$$\frac{\partial \Phi}{\partial x} = -L \frac{\partial I}{\partial t}. \tag{14.12}$$

Differentiating one equation with respect to x and the other with respect to t, and combining so as to eliminate either Φ or I yield a second-order equation either for I

$$\frac{\partial^2 I}{\partial x^2} = LC \frac{\partial^2 I}{\partial t^2} \tag{14.13}$$

or for Φ

$$\frac{\partial^2 \Phi}{\partial x^2} = LC \frac{\partial^2 \Phi}{\partial t^2}. \tag{14.14}$$

Both I and Φ obey the same hyperbolic partial differential equation.

The boundary conditions may be more convenient in one form or the other. For example, the potential at the left end of the line might have a given waveform as a function of t:

$$\Phi(0, t) = V(t). \tag{14.15}$$

At the other end, an "open" line might be specified by setting

$$I = 0 \text{ at } x = x_{max}, \tag{14.16}$$

while a "shorted" line would be specified by setting

$$\Phi(x_{max}, t) = 0. \tag{14.17}$$

Since there is a second derivative with respect to t, both I and $\partial I/\partial t$ should be specified as initial conditions. Alternatively, one could specify both Φ and $\partial\Phi/\partial t$ as initial conditions. As shown by Eqs. 14.11 and 14.12, specifying I as a function of x is equivalent to specifying $\partial\Phi/\partial t$, and specifying Φ as a function of x is equivalent to specifying $\partial I/\partial t$. Thus, the specification of both I and Φ as initial conditions is equivalent to either of the other ways of setting the initial conditions.

As shown by Eq. 14.6, information propagates along characteristic lines with finite velocities. Thus, a certain domain in space is completely specified by the values of I and Φ at an earlier time over a larger domain of x, as determined by the velocity of propagation along the characteristic lines (see Fig. 14.4).

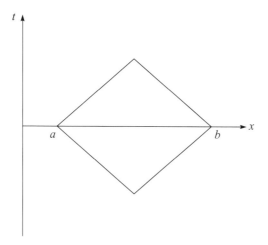

Figure 14.4 Specification of I and Φ at $t = 0$ for $a < x < b$ specifies the solution in the upper triangle.

This can be brought out in another way by transforming the equations in a way suggested by the general solution 14.6. In this case, the velocities of propagation, given by the values of m in Eq. 14.5, are $1/\sqrt{LC}$ and $-1/\sqrt{LC}$, and we define two new coordinates as

$$\alpha = x + t/\sqrt{LC} \text{ and } \beta = x - t/\sqrt{LC}. \tag{14.18}$$

Transformation of Eqs. 14.11 and 14.12 from the independent variables x and t to the new independent variables α and β leads to the equations

$$\frac{\partial I}{\partial \beta} - \sqrt{\frac{C}{L}} \frac{\partial \Phi}{\partial \beta} = 0 \qquad (14.19)$$

and

$$\frac{\partial I}{\partial \alpha} + \sqrt{\frac{C}{L}} \frac{\partial \Phi}{\partial \alpha} = 0. \qquad (14.20)$$

(Coordinate transformations of this type are treated in Chapter 16.) This might lead us to define F and G as

$$F = I - \sqrt{C/L}\,\Phi \qquad (14.21)$$

and

$$G = I + \sqrt{C/L}\,\Phi \qquad (14.22)$$

so that the governing equations are

$$\frac{\partial F}{\partial \beta} = 0 \quad \text{and} \quad \frac{\partial G}{\partial \alpha} = 0 \qquad (14.23)$$

with the solutions

$$F = f(\alpha) \text{ and } G = g(\beta), \qquad (14.24)$$

where f and g are integration "constants." f is independent of β but can be a function of α, and g is independent of α but can be a function of β. Thus, we see that the general solution to these equations is represented by the arbitrary functions f and g: F is constant along lines of constant α, and G is constant along lines of constant β. The values of I and Φ at any given values of x and t are determined by first getting the values of F and G and then resolving these into values of I and Φ. For example,

$$2I = F + G = f(\alpha) + g(\beta), \qquad (14.25)$$

a form of Eq. 14.6.

Chapter 24 treats a hyperbolic problem. Acrivos [1] has provided us with examples of interest in chemical engineering.

Reference

1. Andreas Acrivos. "Method of characteristics technique. Application to heat and mass transfer problems," *Industrial and Engineering Chemistry Process Design and Development*, **48**, 703–710 (1956).

Chapter 15

Steady Heat Conduction in a Rectangle

We intend to treat steady heat conduction in the rectangular region shown in Fig. 15.1. It is assumed that the temperature is independent of the third coordinate z and consequently satisfies Laplace's equation in two dimensions

$$\frac{\partial^2 T}{\partial x^2} + \frac{\partial^2 T}{\partial y^2} = 0. \tag{15.1}$$

For boundary conditions, we choose

$$\left.\begin{array}{l} \partial T/\partial x = 0 \quad \text{at } x = 0 \\ T = 0 \quad \text{at } x = L \\ T = 0 \quad \text{at } y = W \\ T = f(x) \quad \text{at } y = 0 \end{array}\right\}. \tag{15.2}$$

It should be noticed that the equation is elliptic and it is appropriate to specify boundary conditions on all the boundaries of the region being considered.

We first attempt a solution by the *method of separation of variable*, that is, we assume that the solution $T(x, y)$ can be expressed as the product of a function of x and a second function that depends only on y:

$$T(x, y) = X(x)Y(y). \tag{15.3}$$

The Newman Lectures on Mathematics
John Newman and Vincent Battaglia
Copyright © 2018 Pan Stanford Publishing Pte. Ltd.
ISBN 978-981-4774-25-3 (Hardcover), 978-1-315-10885-8 (eBook)
www.panstanford.com

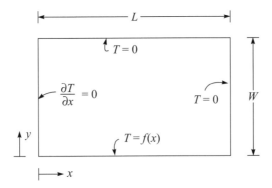

Figure 15.1 Boundary conditions for steady heat conduction in a rectangle.

Substitution into Eq. 15.1 gives

$$Y\frac{d^2X}{dx^2} + X\frac{d^2Y}{dy^2} = 0$$

or

$$\frac{1}{X}\frac{d^2X}{dx^2} = -\frac{1}{Y}\frac{d^2Y}{dy^2} = -\lambda^2. \tag{15.4}$$

We recognize that these terms must be constant since the first term is independent of y and the second term is independent of x.

The function $X(x)$ is thus to be determined from the equation

$$\frac{d^2X}{dx^2} + \lambda^2 X = 0 \tag{15.5}$$

and the boundary conditions

$$dX/dx = 0 \quad \text{at } x = 0 \quad \text{and} \quad X = 0 \quad \text{at } x = L. \tag{15.6}$$

This is a Sturm–Liouville problem, which has solutions only for certain values of λ:

$$\lambda = \lambda_m = \left(n - \frac{1}{2}\right)\pi/L, \; n = 1, 2, \dots. \tag{15.7}$$

For these eigenvalues, we obtain the corresponding eigenfunctions:

$$X = X_n(x) = \cos(\lambda_n x) = \cos\left[\left(n - \frac{1}{2}\right)\pi x/L\right]. \tag{15.8}$$

Of course, the eigenfunctions are determined by Eqs. 15.5 and 15.6, except for a constant factor; here we have made an arbitrary choice for that factor.

The function Y satisfies the equation

$$\frac{d^2Y}{dy^2} = \lambda^2 Y . \tag{15.9}$$

We choose the solution

$$Y = Y_n(y) = \frac{\sin h[\lambda_n(W - y)]}{\sin h(\lambda_n W)} \tag{15.10}$$

which satisfies the condition $Y = 0$ at $y = W$. This time we have adjusted the constant factor so that $Y = 1$ when $y = 0$.

We have now obtained a solution, or a number of solutions, $X_n(x)Y_n(y)$, which satisfies the differential equation 15.1 and the first three boundary conditions, those at $x = 0$, $x = L$, and $y = W$. None of these solutions will, in general, satisfy the remaining boundary condition at $y = 0$. However, since the problem is linear, we can superpose these solutions to obtain a more general solution

$$T(x, y) = \sum_{n=1}^{\infty} C_n X_n(x) Y_n(y) = \sum_{n=1}^{\infty} C_n \cos[\lambda_n x] \frac{\sin h[\lambda_n(W - y)]}{\sin h(\lambda_n W)} ,$$

$$\tag{15.11}$$

with which we can hope to satisfy a more general boundary condition at $y = 0$. Indeed, at $y = 0$, we have

$$T(x, 0) = f(x) = \sum_{n=1}^{\infty} C_n \cos(\lambda_n x). \tag{15.12}$$

Since the functions $X_n(x)$ were obtained from the solution to a Sturm–Liouville problem, we recognize that Eq. 15.12 represents the expansion of the function $f(x)$ in a series of orthogonal functions $X_n(x)$. Furthermore, the functions $X_n(x)$ are complete in the sense that practically any function $f(x)$ can be adequately represented by the series expansion 15.12. Finally, we should note that there was no loss of generality in specifying the arbitrary, constant factors in the solutions 15.8 and 15.10 for $X_n(x)$ and $Y_n(y)$ since we have multiplied the solution so obtained by the coefficient in the superposition solution 15.11.

The values of the coefficients C_n are to be determined from Eq. 15.12 and the orthogonality property of the eigenfunctions $X_n(x)$. Thus, we multiply Eq. 15.12 by $X_k(x)$ and integrate from 0 to L to obtain

$$\int_0^L f(x)\cos(\lambda_k x)dx = C_k \int_0^L \cos^2(\lambda_k x)dx = \frac{1}{2}LC_k \qquad (15.13)$$

or

$$C_n = \frac{2}{L}\int_0^L f(x)\cos(\lambda_n x)dx. \qquad (15.14)$$

This completes the solution to the problem by the method of separation of variables. We have obtained the solution 15.11 with the coefficients determined by Eq. 15.14. We should review what we have done. We have treated a linear, homogeneous, differential equation subject to boundary conditions, all but one of which are also homogeneous. We have assumed a solution by separation of variables in a coordinate system for which the differential equation allows such a solution, and in which coordinate system the boundaries are represented by appropriately simple formulas. The solution by separation of variables then yielded a separation constant λ and a Sturm–Liouville system for one of the functions. Solutions to this Sturm–Liouville system existed only for certain eigenvalues λ_n of the separation constant. By superposition of the solutions thus obtained, we obtained a fairly general solution that satisfied the differential equation and the homogeneous boundary conditions. The one remaining, nonhomogeneous boundary condition could be satisfied by appropriate selection of the coefficients of the superposition solution, and this process of the determination of the coefficients was found to be straightforward because of the orthogonality property of the solutions to the Sturm–Liouville system.

These features, as you may have the opportunity to observe for yourselves, are common to the solution to problems by the method of separation of variables. Conversely, the method may not be so straightforward or may fail completely for a number of reasons: the problem is nonlinear, the coordinate system and the differential equation are not compatible with a separation of variables, the differential equation or more than two boundary conditions are nonhomogeneous, or the boundaries cannot be simply expressed in the coordinate system.

Now let us consider in more detail a special case of the problem treated above; let

$$f(x) = T_0, \tag{15.15}$$

which corresponds to a constant temperature along the boundary $y = 0$. Equation 15.14 yields the coefficients in the series solution 15.11:

$$C_n = \frac{2T_0}{\lambda_n L} \sin(\lambda_n L) = -(-1)^n \frac{2T_0}{\pi\left(n - \dfrac{1}{2}\right)}. \tag{15.16}$$

As engineers, we might be interested in the heat flux at the various boundaries:

$$q_y(x, 0) = -k \frac{\partial T}{\partial y}\bigg|_{y=0} = k \sum_{n=1}^{\infty} C_n \cos(\lambda_n x) \lambda_n \frac{\cosh(\lambda_n W)}{\sinh(\lambda_n W)}$$

$$= \frac{2T_0 k}{L} \sum_{n=1}^{\infty} \frac{\sin(\lambda_n L)\cos(\lambda_n x)}{\tanh(\lambda_n W)} = \frac{2T_0 k}{L} \sum_{n=1}^{\infty} \frac{\sin[\lambda_n(L-x)]}{\tanh(\lambda_n W)}. \tag{15.17}$$

$$q_x(L, y) = -k \frac{\partial T}{\partial x}\bigg|_{x=L} = k \sum_{n=1}^{\infty} C_n \lambda_n \sin(\lambda_n L) \frac{\sinh[\lambda_n(W-y)]}{\sinh(\lambda_n W)}$$

$$= \frac{2T_0 k}{L} \sum_{n=1}^{\infty} \frac{\sinh[\lambda_n(W-y)]}{\sinh(\lambda_n W)}. \tag{15.18}$$

$$q_y(x, W) = -k \frac{\partial T}{\partial y}\bigg|_{y=W} = k \sum_{n=1}^{\infty} C_n \cos(\lambda_n x) \lambda_n \frac{\cosh(0)}{\sinh(\lambda_n W)}$$

$$= \frac{2T_0 k}{L} \sum_{n=1}^{\infty} \frac{\sin[\lambda_n(L-x)]}{\sinh(\lambda_n W)}. \tag{15.19}$$

Down in the lower right corner, we should expect to encounter a singular behavior that would not be revealed clearly by the series solution. Consequently, let us give a separate treatment to that region. If we express the temperature distribution in terms of the variables $L-x$ and y, then in the corner region the solution so expressed should be independent of the parameters L and W. Then it can be shown by dimensional arguments that the dimensionless ratio T/T_0 must depend only on a dimensionless ratio of the variables y and $L-x$, and there is only one independent ratio of this type, $\eta = y/(L-x)$, all other such dimensionless ratios being expressible in terms of η. Thus, our argument is that in the corner region

$$T/T_0 = g(\eta) \quad \text{where} \quad \eta = y/(L-x). \tag{15.20}$$

The method of solution here is an example of a *similarity transformation*, a very important method in the treatment of partial differential equations. Our argument leads us to the conclusion that g must satisfy an ordinary differential equation in which only the variable η appears, but not the variables y and $L-x$, except in this combination. Let us determine this differential equation.

$$
\left.
\begin{aligned}
\frac{1}{T_0}\frac{\partial T}{\partial y} &= \frac{dg}{d\eta}\frac{\partial \eta}{\partial y} = \frac{1}{L-x}\frac{dg}{d\eta} \\[2mm]
\frac{1}{T_0}\frac{\partial^2 T}{\partial y^2} &= \frac{1}{(L-x)^2}\frac{d^2 g}{d\eta^2} \\[2mm]
\frac{1}{T_0}\frac{\partial T}{\partial x} &= \frac{dg}{d\eta}\frac{\partial \eta}{\partial x} = \frac{y}{(L-x)^2}\frac{dg}{d\eta} = \frac{\eta}{L-x}\frac{dg}{d\eta} \\[2mm]
\frac{1}{T_0}\frac{\partial^2 T}{\partial x^2} &= \frac{2y}{(L-x)^3}\frac{dg}{d\eta} + \frac{y}{(L-x)^2}\frac{d^2 g}{d\eta^2}\frac{\partial \eta}{\partial x} = \frac{2\eta}{(L-x)^2}\frac{dg}{dn} + \frac{\eta^2}{(L-x)^2}\frac{d^2 g}{d\eta^2}
\end{aligned}
\right\}.
$$

$$\tag{15.21}$$

Substitution of these results into the partial differential equation 15.1 yields

$$\frac{2\eta}{(L-x)^2}\frac{dg}{d\eta} + \frac{\eta^2}{(L-x)^2}\frac{d^2 g}{d\eta^2} + \frac{1}{(L-x)^2}\frac{d^2 g}{d\eta^2} = 0 \tag{15.22}$$

or

$$\left(1+\eta^2\right)\frac{d^2 g}{d\eta^2} + 2\eta\frac{dg}{d\eta} = 0. \tag{15.23}$$

We notice that Eq. 15.23 is independent of the variables y and $L-x$ except in the combination η. Furthermore, the boundary conditions become

$$g = 1 \quad \text{at} \quad \eta = 0 \quad \text{and} \quad g = 0 \quad \text{at} \quad \eta = \infty \tag{15.24}$$

and are also independent of y and $L-x$ separately. Equation 15.23 is a Legendre equation of imaginary argument of order zero and has the solution

$$g(\eta) = 1 - \frac{2}{\pi} \tan^{-1}\eta \qquad (15.25)$$

giving

$$\frac{dg}{d\eta} = -\frac{2}{\pi}\frac{1}{1+\eta^2}. \qquad (15.26)$$

We now have an explicit expression for the behavior of the solution in the corner region and can calculate the heat flux at the boundaries of the rectangle near the corner:

$$q_y(x,0) = -k\frac{\partial T}{\partial y}\bigg|_{y=0} \rightarrow -\frac{kT_0}{L-x}\frac{dg}{d\eta}\bigg|_{\eta=0} = \frac{2}{\pi}\frac{kT_0}{L-x} \text{ as } x \rightarrow L.$$

$$(15.27)$$

$$q_x(L,y) = -k\frac{\partial T}{\partial x}\bigg|_{x=L} \rightarrow \lim_{\eta\rightarrow\infty} -\frac{kT_0\eta}{L-x}\frac{dg}{d\eta} = \frac{2}{\pi}\frac{kT_0}{y} \text{ as } y \rightarrow 0.$$

$$(15.28)$$

We see that not only does the heat flux become infinite in the corner region but also the heat flux cannot be integrated and the total heat transfer is infinite. Consequently, we should not expect to encounter such an unrealistic problem in practice.

Now let us return to the series solutions, for which Eqs. 15.17 to 15.19 express the heat flux at the boundaries. Unfortunately, Eq. 15.17 does not converge for any values of x and consequently gives us no information on the behavior of the heat flux on the lower boundary. Equations 15.18 and 15.19 do give useful results. At this point one might wonder what to do. We could choose another example, which is better behaved, but we decide to leave this to the exercises. Instead, we forge ahead and salvage what we can.

To facilitate our calculations and to eliminate a parameter, we confine ourselves to the case of $W/L \rightarrow \infty$. Then Eqs. 15.10, 15.11, and 15.17 to 15.19 become

$$Y_n(y) = e^{-\lambda_n y} \qquad (15.29)$$

$$T(x, y) = \sum_{n=1}^{\infty} C_n \cos(\lambda_n x) e^{-\lambda_n y} \qquad (15.30)$$

$$q_y(x, 0) = \frac{2T_0 k}{L} \sum_{n=1}^{\infty} \sin[\lambda_n(L - x)] \qquad (15.31)$$

$$q_x(L, y) = \frac{2T_0 k}{L} \sum_{n=1}^{\infty} e^{-\lambda_n y} \qquad (15.32)$$

$$q_y(x, W) = \frac{4T_0 k}{L} \sum_{n=1}^{\infty} \sin[\lambda_n(L - x)] e^{-\lambda_n W} \to 0. \qquad (15.33)$$

Again, Eq. 15.31 does not converge. Figure 15.2 represents the heat flux on the right boundary as given by Eq. 15.32. The flux is multiplied by $\pi y / 2kT_0$ so that the behavior near $y = 0$ can be compared with that predicted by Eq. 15.28, which is supposed to be valid in the corner region. The agreement is excellent. We also see from Fig. 15.2 that the heat flux drops off rapidly for $y > L$, and consequently the heat flux at the upper boundary, given by Eq. 15.33, would be of little interest.

Figure 15.3 shows the same thing as Fig. 15.2 but illustrates the result of truncating the series 15.32 after a finite number of terms. Good accuracy can be obtained for any value of y by carrying enough terms, but for a given number of terms, the accuracy is lost for small values of y. This illustrates the value of the asymptotic solution for the corner region, as given by Eq. 15.28, which gives us definite information on the behavior of the heat flux near $y = 0$, information which could not be obtained with certainty from the series solution. It also permits us to avoid the folly of trying to integrate the heat flux given by the series solution in order to obtain the total rate of heat transfer. At the same time, we are confident that Eq. 15.27 gives us information about the behavior of the heat flux on the lower boundary, where the series solution failed to give us any information.

We are still faced with the problem of obtaining a solution that gives the heat flux on the lower boundary. Two avenues of approach are open to us: the use of conformal mapping and the use of Laplace transforms. Both methods should be more tractable if we still regard W/L to be infinite. The conformal mapping is more likely to yield a useful answer, but the method is not developed until Chapter 25.

The Laplace transforms would be nice to illustrate a new technique, if they could be made to work, but they are better suited for initial value problems. In the present case, we do not know both the temperature and the heat flux on any of the boundaries; in fact this is what we seek.

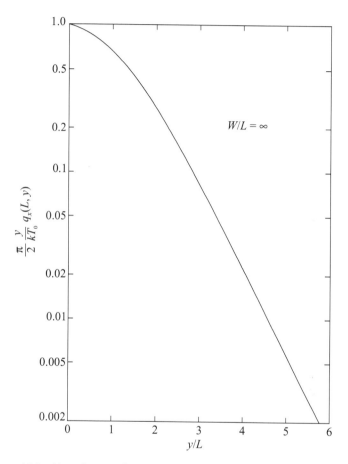

Figure 15.2 Heat flux on the right boundary, multiplied by y and constant parameters.

By conformal mapping, we obtain the solution for the temperature distribution (with $W/L = \infty$)

$$T = T_0 \left[1 - \frac{2}{\pi} \tan^{-1} \left(\frac{\sinh(\pi y/2L)}{\cos(\pi x/2L)} \right) \right]. \qquad (15.34)$$

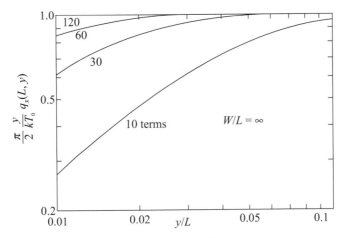

Figure 15.3 Heat flux on the right boundary, illustrating the result of carrying various number of terms.

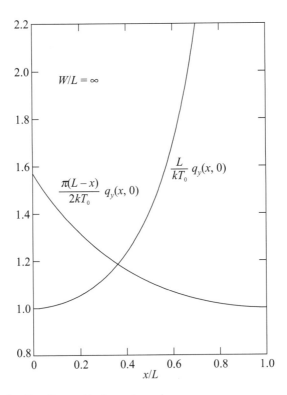

Figure 15.4 Heat flux on the lower boundary.

Consequently, the expressions for the heat fluxes at the boundaries are

$$q_y(x, 0) = kT_0/L \cos(\pi x/2L) \qquad (15.35)$$

$$q_x(L, y) = kT_0/L \sinh(\pi y/2L). \qquad (15.36)$$

Equation 15.36 agrees with Fig. 15.2. The heat flux on the lower boundary, given by Eq. 15.35, is plotted in Fig. 15.4. It is multiplied by $\pi(L-x)/2kT_0$ so that the behavior near $x = L$ can be compared with that predicted by Eq. 15.27.

Problems

15.1 Rework the problem of steady heat conduction in a rectangle with

$$f(x) = T_0\left(1 - \frac{x}{L}\right).$$

In this case, the temperature will be continuous on the boundary.

(a) Calculate the coefficients in the series solution for the temperature distribution as given by the method of separation of variables.

(b) Obtain series expressions for the normal heat fluxes at the lower boundary and the right boundary. Determine whether or not these series converge.

(c) Treat the region in the lower right corner ($x \to L$ and $y \to 0$). Show by *dimensional arguments* that in this region the solution must be of the form

$$T = (T_0/L)(L-x)g\left(\frac{y}{L-x}\right).$$

Note that in this region the temperature distribution depends on the parameter T_0/L but not on T_0 separately. Determine the function g, and from this result, find asymptotic expressions (valid in the corner region) for the heat fluxes at the boundaries. If possible, compare these with the series expressions obtained in part b.

15.2 For $f(x) = T_0$, verify that the temperature distribution when $W/L = \infty$ is, indeed, given by

$$T = T_0 \left[1 - \frac{2}{\pi} \tan^{-1} \left(\frac{\sin h(\pi y/2L)}{\cos(\pi x/2L)} \right) \right].$$

From this result, obtain expressions for the heat fluxes at the lower and right boundaries.

15.3 Treat steady heat conduction in a circular region, or a cylinder with no temperature variation in the direction along the axis. The temperature distribution on the perimeter is $f(\theta)$. Use separation of variables and show where the Sturm–Liouville system arises, which limits the separation constant to certain eigenvalues. Determine the functions of the two variables in the product solution. Superpose solutions corresponding to different eigenvalues and show how to determine the coefficients.

Chapter 16

Coordinate Transformations

Frequently, a problem stated in terms of one set of independent variables or coordinates needs to be restated in terms of a different but equivalent set of independent variables. It then may be necessary to find relationships among partial derivatives involving the two sets of variables. Let the old set of variables be $x, y, z, \ldots$ and let the new set be $r, s, t, \ldots$. The basis for the desired relationships is the formula for the total differential:

$$dT = \frac{\partial T}{\partial x}dx + \frac{\partial T}{\partial y}dy + \frac{\partial T}{\partial z}dz + \cdots. \qquad (16.1)$$

This may then be used to obtain partial derivatives with respect to the new variables, for example,

$$\frac{\partial T}{\partial r} = \frac{\partial T}{\partial x}\frac{\partial x}{\partial r} + \frac{\partial T}{\partial y}\frac{\partial y}{\partial r} + \frac{\partial T}{\partial z}\frac{\partial z}{\partial r} + \cdots. \qquad (16.2)$$

However, in using this formula, care must be exercised with regard to which variables are held constant in each derivative. As an aid in this endeavor, we can indicate explicitly the variables that are held constant. Equation 16.1 would then be written as

$$dT = \left(\frac{\partial T}{\partial x}\right)_{y,z,\ldots} dx + \left(\frac{\partial T}{\partial y}\right)_{x,z,\ldots} dy + \left(\frac{\partial T}{\partial z}\right)_{x,y,\ldots} dz + \cdots. \qquad (16.3)$$

The Newman Lectures on Mathematics
John Newman and Vincent Battaglia
Copyright © 2018 Pan Stanford Publishing Pte. Ltd.
ISBN 978-981-4774-25-3 (Hardcover), 978-1-315-10885-8 (eBook)
www.panstanford.com

The original intent in writing the derivative with respect to r in Eq. 15.2 was that the variables $s, t, \ldots$ should be fixed:

$$\left(\frac{\partial T}{\partial r}\right)_{s,t,\ldots} = \left(\frac{\partial T}{\partial x}\right)_{y,z,\ldots} \left(\frac{\partial x}{\partial r}\right)_{s,t,\ldots} + \left(\frac{\partial T}{\partial y}\right)_{x,z,\ldots} \left(\frac{\partial y}{\partial r}\right)_{s,t,\ldots}$$

$$+ \left(\frac{\partial T}{\partial z}\right)_{x,y,\ldots} \left(\frac{\partial z}{\partial r}\right)_{s,t,\ldots} + \cdots . \qquad (16.4)$$

Mathematicians will tell you that when each independent variable belongs to a definite set of independent variables, such subscripts are superfluous; differentiation with respect to one variable automatically tells you that all other variables of that set are to be held constant.

But in thermodynamics, the possibility for confusion must baffle even the mathematicians since here eight state functions may be said to depend on any two of the eight (for a one-component system).

Furthermore, one should be careful to note that the variables $r, s, t, \ldots$ may be explicitly stated in terms of $x, y, z, \ldots$ or vice versa. Thus, while

$$(\partial x/\partial r)_{s,t,\ldots} = 1/(\partial r/\partial x)_{s,t,\ldots}, \qquad (16.5)$$

in general, we find that

$$(\partial x/\partial r)_{s,t,\ldots} \neq 1/(\partial r/\partial x)_{y,z,\ldots}. \qquad (16.6)$$

In cylindrical coordinates, the Laplacian is

$$\nabla^2 \Phi = \frac{1}{r}\frac{\partial}{\partial r}\left(r\frac{\partial \Phi}{\partial r}\right) + \frac{1}{r^2}\frac{\partial^2 \Phi}{\partial \theta^2} + \frac{\partial^2 \Phi}{\partial z^2}. \qquad (16.7)$$

Let us change this to rotational elliptic or oblate spheroidal coordinates η, ξ, θ defined by

$$\theta = \theta, \quad z = r_0 \xi \eta, \quad r = r_0 \sqrt{(1+\xi^2)(1-\eta^2)}. \qquad (16.8)$$

The coordinates are illustrated in Fig. 16.1 and find application in problems where the geometry is suitable, as illustrated in the next two chapters.

In this example, r, θ, and z are expressed in terms of η, ξ, and θ, and the inversion is possible but complicated. Hence, we start by expressing derivatives with respect to η, ξ, and θ in terms of derivatives with respect to r, θ, and z, even though we will need to invert the results.

$$\left(\frac{\partial\Phi}{\partial\eta}\right)_{\xi,\theta} = \left(\frac{\partial\Phi}{\partial r}\right)_{\theta,z}\left(\frac{\partial r}{\partial\eta}\right)_{\xi,\theta} + \left(\frac{\partial\Phi}{\partial\theta}\right)_{r,z}\left(\frac{\partial\theta}{\partial\eta}\right)_{\xi,\theta} + \left(\frac{\partial\Phi}{\partial z}\right)_{r,\theta}\left(\frac{\partial z}{\partial\eta}\right)_{\xi,\theta}$$

$$= -r_0\eta\frac{\sqrt{1+\xi^2}}{\sqrt{1-\eta^2}}\frac{\partial\Phi}{\partial r} + r_0\xi\frac{\partial\Phi}{\partial z}. \tag{16.9}$$

$$\left(\frac{\partial\Phi}{\partial\xi}\right)_{\eta,\theta} = r_0\xi\frac{\sqrt{1-\eta^2}}{\sqrt{1+\xi^2}}\frac{\partial\Phi}{\partial r} + r_0\eta\frac{\partial\Phi}{\partial z}. \tag{16.10}$$

$$\left(\frac{\partial\Phi}{\partial\theta}\right)_{\eta,\xi} = \left(\frac{\partial\Phi}{\partial\theta}\right)_{r,z}. \tag{16.11}$$

Inverting Eqs. 16.9 and 16.10, we get

$$\frac{\partial\Phi}{\partial r} = \frac{r}{r_0^2(\eta^2+\xi^2)}\left[\xi\frac{\partial\Phi}{\partial\xi} - \eta\frac{\partial\Phi}{\partial\eta}\right] \tag{16.12}$$

$$\frac{\partial\Phi}{\partial z} = \frac{1}{r_0(\eta^2+\xi^2)}\left[(1+\xi^2)\eta\frac{\partial\Phi}{\partial\xi} + (1-\eta^2)\xi\frac{\partial\Phi}{\partial\eta}\right]. \tag{16.13}$$

Since the Laplacian contains second derivatives, it is necessary now to differentiate Eqs. 16.9 to 16.11. This is a bit tricky but yields

$$\frac{\partial^2\Phi}{\partial\eta^2} = r_0^2\eta^2\frac{1+\xi^2}{1-\eta^2}\frac{\partial^2\Phi}{\partial r^2} + r_0^2\xi^2\frac{\partial^2\Phi}{\partial z^2} - r_0\frac{\sqrt{1+\xi^2}}{(1-\eta^2)^{3/2}}\frac{\partial\Phi}{\partial r}$$

$$- 2r_0^2\eta\xi\frac{\sqrt{1+\xi^2}}{\sqrt{1-\eta^2}}\frac{\partial^2\Phi}{\partial r\partial z}. \tag{16.14}$$

$$\frac{\partial^2\Phi}{\partial\xi^2} = r_0^2\xi^2\frac{1-\eta^2}{1+\xi^2}\frac{\partial^2\Phi}{\partial r^2} + r_0^2\eta^2\frac{\partial^2\Phi}{\partial z^2} + r_0\frac{\sqrt{1-\eta^2}}{(1+\xi^2)^{3/2}}\frac{\partial\Phi}{\partial r}$$

$$+ 2r_0^2\eta\xi\frac{\sqrt{1-\eta^2}}{\sqrt{1+\xi^2}}\frac{\partial^2\Phi}{\partial r\partial z}. \tag{16.15}$$

$$\frac{\partial^2\Phi}{\partial\theta^2} = \frac{\partial^2\Phi}{\partial\theta^2}. \tag{16.16}$$

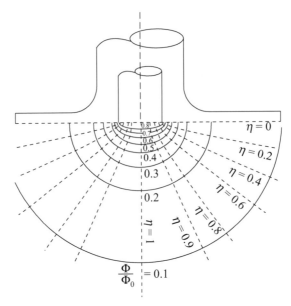

Figure 16.1   Rotational elliptic coordinates η and ξ for an axisymmetric geometry. Here ξ is given as $\dfrac{\Phi}{\Phi_0} = 1 - \dfrac{2}{\pi}\tan^{-1}\xi$.

Now we can take a linear combination of Eqs. 16.14 and 16.15 in order to eliminate the cross derivative $\partial^2\Phi/\partial r\partial z$. This is accomplished by multiplying the second equation by $(1+\xi^2)/(1-\eta^2)$ and adding to the preceding equation:

$$\frac{\partial^2\Phi}{\partial\eta^2} + \frac{1+\xi^2}{1-\eta^2}\frac{\partial^2\Phi}{\partial\xi^2} = r_0^2\frac{\xi^2+\eta^2}{1-\eta^2}\left[\frac{\partial^2\Phi}{\partial r^2} + \frac{\partial^2\Phi}{\partial z^2} - \frac{1}{r}\frac{\partial\Phi}{\partial r}\right]. \qquad (16.17)$$

Since in Eq. 16.17 the coefficients of $\partial^2\Phi/\partial r^2$ and $\partial^2\Phi/\partial z^2$ are equal, we can now combine this equation with Eqs. 16.12 and 16.16 to obtain an expression for the Laplacian in terms of rotational elliptic coordinates:

$$\nabla^2\Phi = \frac{1}{r_0^2}\left\{\frac{1}{\xi^2+\eta^2}\left[(1-\eta^2)\frac{\partial^2\Phi}{\partial\eta^2} - 2\eta\frac{\partial\Phi}{\partial\eta} + (1+\xi^2)\frac{\partial^2\Phi}{\partial\xi^2} + 2\xi\frac{\partial\Phi}{\partial\xi}\right]\right.$$

$$\left. + \frac{1}{(1+\xi^2)(1-\eta^2)}\frac{\partial^2\Phi}{\partial\theta^2}\right\}. \qquad (16.18)$$

Your textbook may treat curvilinear coordinates in a general and abstract manner, which we shall not repeat in the chapter. We shall restrict ourselves to rectangular, cylindrical, spherical, and rotational elliptic coordinates. Moon and Spencer [1] summarize a wealth of information on other coordinate systems that is useful in physical problems.

Reference

1. Parry Moon and Domina Eberle Spencer. *Field Theory Handbook.* Berlin: Springer Verlag, 1961.

Chapter 17

Disk Electrode in an Insulating Plane

A disk electrode of radius r_0 is embedded in a large, insulating plane, and the counterelectrode is at infinity. The electrostatic potential in an electrolytic solution of uniform composition satisfies Laplace's equation

$$\nabla^2 \Phi = 0. \tag{17.1}$$

With neglect of the limitations of electrode kinetics and of concentration variations near the electrode, the potential adjacent to the disk can be assumed to be constant:

$$\Phi = \Phi_0 \quad \text{at} \quad z = 0, r \leq r_0, \tag{17.2}$$

while the potential at infinity may be taken to be zero:

$$\Phi \to 0 \quad \text{as} \quad r^2 + z^2 \to \infty. \tag{17.3}$$

On the insulating plane, the current density is zero, which implies

$$\frac{\partial \Phi}{\partial z} = 0 \quad \text{at} \quad z = 0, r > r_0. \tag{17.4}$$

One could make the mistake of seeking a solution by separation of variables in spherical coordinates ρ, ϑ, and φ, where the Laplacian is

$$\frac{1}{\rho^2} \frac{\partial}{\partial \rho}\left(\rho^2 \frac{\partial \Phi}{\partial \rho}\right) + \frac{1}{\rho^2 \sin\varphi} \frac{\partial}{\partial \varphi}\left(\sin\varphi \frac{\partial \Phi}{\partial \varphi}\right) + \frac{1}{\rho^2 \sin^2\varphi} \frac{\partial^2 \Phi}{\partial \vartheta^2} = 0. \tag{17.5}$$

The Newman Lectures on Mathematics
John Newman and Vincent Battaglia
Copyright © 2018 Pan Stanford Publishing Pte. Ltd.
ISBN 978-981-4774-25-3 (Hardcover), 978-1-315-10885-8 (eBook)
www.panstanford.com

Since the problem is axisymmetric, we let

$$\Phi = R(\rho)F(\varphi). \tag{17.6}$$

Equation 17.5 becomes

$$\frac{1}{R}\frac{d}{d\rho}\left(\rho^2\frac{dR}{d\rho}\right) = -\frac{1}{F\sin\varphi}\frac{d}{d\varphi}\left(\sin\varphi\frac{dF}{d\varphi}\right) = p(p+1). \tag{17.7}$$

where the separation constant has been written as $p(p+1)$. F thus satisfies the equation

$$\frac{1}{\sin\varphi}\frac{d}{d\varphi}\left(\sin\varphi\frac{dF}{d\varphi}\right) + p(p+1)F = 0, \tag{17.8}$$

while R satisfies the Euler equation

$$\rho^2\frac{d^2R}{d\rho^2} + 2\rho\frac{dR}{d\rho} - p(p+1)R = 0, \tag{17.9}$$

with the solution

$$R = A\rho^p + B\rho^{-p-1}. \tag{17.10}$$

The solution for R has the characteristic that it is bounded at both $\rho = 0$ and $\rho = \infty$ only if $p = 0$ or $p = -1$ (or both A and B are zero). Furthermore, there will be problems in satisfying the boundary conditions on the plane $z = 0$. Therefore, we consider separately the region $\rho < r_0$, and the region $\rho > r_0$, hoping eventually to match at $\rho = r_0$ the solutions obtained in the two regions. Note that we can restrict p to be greater than $-1/2$ with no loss in generality.

Consider first the region $\rho > r_0$. Then F satisfies the boundary conditions

$$\left.\begin{array}{l} dF/d\varphi = 0 \quad\text{at}\quad \varphi = \pi/2 \\ F \text{ is well behaved at } \varphi = 0 \end{array}\right\}. \tag{17.11}$$

With the new variable $\mu = \cos\varphi$, we find that F satisfies Legendre's equation

$$(1-\mu^2)\frac{d^2F}{d\mu^2} - 2\mu\frac{dF}{d\mu} + p(p+1)F = 0 \tag{17.12}$$

with the boundary conditions

$$\left.\begin{array}{l} dF/d\mu = 0 \quad\text{at}\quad \mu = 0 \\ F \text{ is well behaved at } \mu = 1 \end{array}\right\}. \tag{17.13}$$

To satisfy the first boundary condition, we restrict ourselves to the first series solution in Eqs. 17.9 to 17.16. The second boundary condition requires that the series terminates, which in turn requires that p be an even integer

$$p = 2n, n = 0, 1, ...\tag{17.14}$$

Thus, the solution for F can be written

$$F = P_p(\mu) = P_{2n}(\mu) = P_{2n}(\cos\varphi),$$

that is, the Legendre polynomial of $\cos\varphi$. On the other hand, in the solution for R in Eq. 17.10 we must set A equal to zero in order to satisfy the boundary condition $R = 0$ at $\rho = \infty$.

$$R = B\rho^{-2n-1}.\tag{17.15}$$

Finally, we can write the general solution in the region $\rho > r_0$ as a superposition of the solutions obtained here:

$$\Phi = \sum_{n=0}^{\infty} B_n P_{2n}(\cos\varphi)\rho^{-2n-1}, \quad \rho > r_0.\tag{17.16}$$

Now let us turn our attention to the region $\rho < r_0$. Here we write the solution as

$$\Phi = \Phi_0 + \Phi_1,\tag{17.17}$$

where Φ_0 is the boundary value and Φ_1 satisfies Laplace's equation and the boundary conditions

$$\left.\begin{array}{l}\Phi_1 = 0 \;\; \text{at} \;\; \varphi = \pi/2 \\ \Phi_1 \;\text{is well behaved at}\; \varphi = 0\end{array}\right\}.\tag{17.18}$$

We try again a solution by separation of variables:

$$\Phi_1 = R(\rho)F(\varphi)\tag{17.19}$$

where R satisfies Eq. 17.9 with the solution 17.10 and F satisfies Eq. 17.8 or, in terms of the variable μ, Eq. 17.12. The boundary conditions for F are now

$$\left.\begin{array}{l}F = 0 \;\; \text{at} \;\; \mu = 0 \\ F \;\text{is well behaved at}\; \mu = 1\end{array}\right\}.\tag{17.20}$$

To satisfy the first boundary condition, we select the series solution 8.17. To satisfy the second boundary condition, the series solution must terminate, which means that p must be an odd integer:

$$p = 2n + 1, n = 0, 1, \dots . \tag{17.21}$$

Then the solution for F can be written as the Legendre polynomial

$$F = P_p(\mu) = P_{2n+1}(\mu) = P_{2n+1}(\cos\varphi). \tag{17.22}$$

In the solution for R in Eq. 17.10, we must set $B = 0$ in order that the solution be well behaved at $\rho = 0$.

$$R = A\rho^{2n+1}. \tag{17.23}$$

Thus, the solution in the region $\rho < r_0$ can be written as a superposition of these solutions

$$\Phi = \Phi_0 + \sum_{n=0}^{\infty} A_n P_{2n+1}(\cos\varphi)\rho^{2n+1}, \quad \rho < r_0. \tag{17.24}$$

This brings us to the problem of selecting the coefficients A_n and B_n in Eqs. 17.16 and 17.24 so that these solutions and their derivatives match at $\rho = r_0$. This gives us the two equations

$$\Phi_0 + \sum_{n=0}^{\infty} A_n P_{2n+1}(\cos\varphi)r_0^{2n+1} = \sum_{n=0}^{\infty} B_n P_{2n}(\cos\varphi)r_0^{-2n-1}. \tag{17.25}$$

$$\sum_{n=0}^{\infty}(2n+1)A_n P_{2n+1}(\cos\varphi)r_0^{2n} = -\sum_{n=0}^{\infty}(2n+1)B_n P_{2n}(\cos\varphi)r_0^{-2n-2}.$$

$$\tag{17.26}$$

Although Legendre polynomials satisfy various orthogonality relations, none seem to help to determine A_n and B_n. Thus, after all this work, we reach a dead end.

Consequently, we turn to the solution to the problem in rotational elliptic coordinates, which now seem more appropriate for the geometry than spherical coordinates. From the preceding chapter, we see that axially symmetric solutions to Laplace's equation $\nabla^2\Phi = 0$ satisfy the equation

$$(1-\eta^2)\frac{\partial^2\Phi}{\partial\eta^2} - 2\eta\frac{\partial\Phi}{\partial\eta} + (1+\xi^2)\frac{\partial^2\Phi}{\partial\xi^2} + 2\xi\frac{\partial\Phi}{\partial\xi} = 0. \tag{17.27}$$

A solution by the separation of variables should be possible. Try

$$\Phi = H(\eta) \,\Xi\,(\xi). \tag{17.28}$$

We then have

$$\frac{(1+\xi^2)}{\Xi}\frac{d^2\Xi}{d\xi^2} + \frac{2\xi}{\Xi}\frac{d\Xi}{d\xi} = -\frac{(1-\eta^2)}{H}\frac{d^2H}{d\eta^2} + \frac{2\eta}{H}\frac{dH}{d\eta} = p(p+1),$$

(17.29)

where the separation constant has been denoted as $p(p+1)$ in the anticipation of the fact that H now satisfies Legendre's equation

$$(1-\eta^2)\frac{d^2H}{d\eta^2} - 2\eta\frac{2H}{d\eta} + p(p+1)H = 0.$$

(17.30)

On the other hand, Ξ satisfies the equation

$$(1+\xi^2)\frac{d^2\Xi}{d\xi^2} + 2\xi\frac{d\Xi}{d\xi} - p(p+1)\Xi = 0.$$

(17.31)

The substitution $x = i\xi$ gives the equation

$$(1-x^2)\frac{d^2\Xi}{dx^2} - 2x\frac{d\Xi}{dx} + p(p+1)\Xi = 0$$

(17.32)

and shows that the solution for Ξ will involve Legendre functions of imaginary argument.

In the new coordinate system, the boundary conditions 17.2 to 17.4 become

$$\left.\begin{array}{l} \Phi = \Phi_0 \quad \text{at} \quad \xi = 0 \quad \text{(on the disk electrode)} \\ \partial\Phi/\partial\eta = 0 \quad \text{at} \quad \eta = 0 \quad \text{(on the insulating annulus)} \\ \Phi \to 0 \quad \text{as} \quad \xi \to \infty \quad \text{(far from the disk)} \\ \Phi \text{ is well behaved at } \eta = 1 \text{ (on the axis)} \end{array}\right\}.$$

(17.33)

Consequently, the boundary conditions for H become

$$\left.\begin{array}{l} dH/d\eta = 0 \quad \text{at} \quad \eta = 0 \\ H \text{ is well behaved at } \eta = 1 \end{array}\right\}.$$

(17.34)

The first of these boundary conditions restricts us to series solutions of the form 8.16. This series is finite at $\eta = 1$ if and only if the series terminates, which in turn requires p to be an even integer

$$p = 2n, n = 0, 1, 2, \dots.$$

(17.35)

Thus $H(\eta)$ is the Legendre polynomial

$$H(\eta) = P_{2n}(\eta).$$

(17.36)

The boundary conditions imposed on Ξ in Eq. 17.31 can be taken to be

$$\Xi = 1 \text{ at } \xi = 0 \text{ and } \Xi \to 0 \text{ as } \xi \to \infty. \qquad (17.37)$$

Although we know that the solution can be written in terms of Legendre functions of imaginary argument, we can profitably postpone the explicit expression of that result.

We can now superpose solutions of this type and express the solution as

$$\Phi = \sum_{n=0}^{\infty} C_n P_{2n}(\eta) \Xi_{2n}(\xi). \qquad (17.38)$$

The one remaining boundary condition to be satisfied is

$$\Phi = \Phi_0 \text{ at } \xi = 0 \qquad (17.39)$$

or

$$\Phi_0 = \sum_{n=0}^{\infty} C_n P_{2n}(\eta). \qquad (17.40)$$

Multiplying by $P_{2m}(n)$ and integrating from $\eta = 0$ to $\eta = 1$ yield

$$\Phi_0 \int_0^1 P_{2m}(\eta) d\eta = C_m \int_0^1 [P_{2m}(\eta)]^2 d\eta \qquad (17.41)$$

since these Legendre functions are orthogonal in the sense that

$$\int_0^1 P_{2m}(\eta) P_{2n}(\eta) d\eta = 0 \quad \text{unless } m = n. \qquad (17.42)$$

Furthermore, since $1 = P_0(\eta)$ is also a member of this family, the left side of Eq. 17.41 vanishes unless $m = 0$. In this manner, we obtain for the coefficients C_n

$$C_0 = \Phi_0, C_n = 0, n = 1, 2, \dots . \qquad (17.43)$$

Thus, we see that we need to solve Eq. 17.31 for Ξ only when $p = 2n = 0$. In this case, reduction in order is possible, and we obtain the solution

$$\Xi_0 = 1 - \frac{2}{\pi} \tan^{-1} \xi = \frac{2}{\pi} \text{ctn}^{-1} \xi. \qquad (17.44)$$

The final solution for the potential, therefore, is

$$\Phi = C_0 P_0(\eta)\Xi_0(\xi) = \Phi_0 \frac{2}{\pi} \operatorname{ctn}^{-1}\xi. \qquad (17.45)$$

Equipotential curves are plotted for this problem on Fig 16.1, where they also coincide with curves of constant ξ.

The current density is related to the gradient of the potential by

$$i = -\kappa\nabla\Phi, \qquad (17.46)$$

where κ is the electric conductivity of the solution. In this case, the current density distribution on the disk electrode is of interest and is obtained from the derivative at $z = 0$ or $\xi = 0$:

$$\frac{i_z}{\kappa} = -\frac{\partial\Phi}{\partial z}\bigg|_{z=0} = \frac{-1}{r_0\eta}\frac{\partial\Phi}{\partial\xi}\bigg|_{\xi=0} \qquad (17.47)$$

(see Eq. 16.13). Evaluation from Eq. 17.45 gives

$$i_z = \frac{\kappa}{r_0\eta}\frac{2}{\pi}\Phi_0 = \frac{2\kappa\Phi_0}{\pi\sqrt{r_0^2 - r^2}} \qquad \text{at} \quad z = 0. \qquad (17.48)$$

The total current flowing from the disk is

$$I = \int_0^{r_0} i_z 2\pi r\, dr = 4\kappa r_0\Phi_0. \qquad (17.49)$$

Hence, the resistance (to a counterelectrode at infinity) is $1/4\ \kappa r_0$.

Reference

1. John Newman. "Resistance for flow of current to a disk." *Journal of the Electrochemical Society*, **113**, 501–502 (1966).

Problem

17.1 Now that you have the solution for the potential near a disk electrode, you might try to express it in the form of Eqs. 17.16 and 17.24. Even without doing that you should be able to appreciate that the solution obtained here is far simpler.

Chapter 18

Suspension of Charged Drops

In the study of charged drops [1], one wants to suspend them while making observations during evaporation, and one wants to measure their mass and charge. One device for suspending them involves the imposition of an alternating-current voltage between a plate with a hole in it and two other plates equally spaced above and below the first plate. In addition there may be a constant voltage between the outside two plates to counterbalance the force of gravity. Actually one might think that the constant field would be sufficient, but it allows the drop to wander, providing no stable suspension point, and stray fields will cause the drop to move out of the field of observation. We shall not treat here the movement of the drop but only the calculation of the field in which the drop moves. To do this, we shall move the outside plates to infinity, so that the fields become uniform at $z = \pm\infty$. Furthermore, we shall assume that the alternating frequency is so low that the problem becomes one of electrostatics; magnetic fields can be neglected, and the potential satisfies Laplace's equation

$$\nabla^2 \Phi = 0. \qquad (18.1)$$

The situation is sketched in Fig. 18.1. Obviously, rotational elliptic coordinates are appropriate for this problem. But first we should handle the fields at infinity. Let

$$\Phi = \Phi_1 + E_c z . \qquad (18.2)$$

The Newman Lectures on Mathematics
John Newman and Vincent Battaglia
Copyright © 2018 Pan Stanford Publishing Pte. Ltd.
ISBN 978-981-4774-25-3 (Hardcover), 978-1-315-10885-8 (eBook)
www.panstanford.com

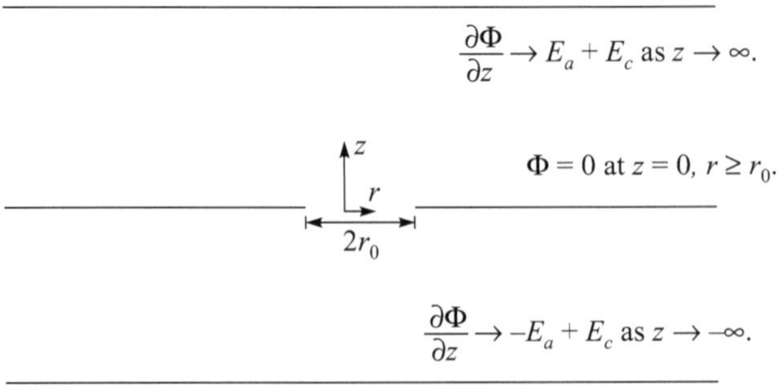

$$\frac{\partial \Phi}{\partial z} \to E_a + E_c \text{ as } z \to \infty.$$

$$\Phi = 0 \text{ at } z = 0, \ r \geq r_0.$$

$$\frac{\partial \Phi}{\partial z} \to -E_a + E_c \text{ as } z \to -\infty.$$

Figure 18.1 Geometric arrangement for drop suspension.

Then Φ_1 satisfies Laplace's equation and the boundary conditions

$$\left.\begin{array}{l} \dfrac{\partial \Phi_1}{\partial z} \to E_a \quad \text{as} \quad z \to \infty \\[2mm] \dfrac{\partial \Phi_1}{\partial z} \to -E_a \quad \text{as} \quad z \to -\infty \\[2mm] \Phi_1 = 0 \quad \text{at} \quad z = 0, \ r \geq r_0 \end{array}\right\}. \qquad (18.3)$$

We also see that by symmetry

$$\partial \Phi_1/\partial z = 0 \quad \text{at} \quad z = 0, r < r_0. \qquad (18.4)$$

This allows us to confine our attention to the upper half-space $(z > 0)$. Now let

$$\Phi_1 = \Phi_2 + E_a|z| \qquad (18.5)$$

Then for the upper half-space, Φ_2 satisfies Laplace's equation and the boundary conditions

$$\left.\begin{array}{l} \Phi_2 = 0 \quad \text{at} \quad z = 0, r \geq r_0 \\[2mm] \Phi_2 \to 0 \quad \text{as} \quad r^2 + z^2 \to \infty \\[2mm] \partial \Phi_2/\partial z = -E_a \quad \text{at} \quad z = 0, r \leq r_0 \end{array}\right\}. \qquad (18.6)$$

Since the problem is axisymmetric, Laplace's equation for Φ_2 in rotational elliptic coordinates reads

$$(1-\eta^2)\frac{\partial^2 \Phi_2}{\partial \eta^2} - 2\eta\frac{\partial \Phi_2}{\partial \eta} + (1+\xi^2)\frac{\partial^2 \Phi_2}{\partial \xi^2} + 2\xi\frac{\partial \Phi_2}{\partial \xi} = 0. \qquad (18.7)$$

The boundary conditions can be written (with the aid of Eq. 16.13) as

$$\left.\begin{array}{l} \Phi_2 = 0 \quad \text{at} \quad \eta = 0 \\ \Phi_2 \to 0 \quad \text{as} \quad \xi \to \infty \\ \partial \Phi_2 / \partial \xi = -r_0 E_a \eta \quad \text{at} \; \xi = 0 \\ \Phi_2 \text{ is well behaved at } \eta = 1 \end{array}\right\}. \qquad (18.8)$$

We seek now a solution by separation of variables

$$\Phi_2 = H(\eta) \, \Xi \, (\xi), \qquad (18.9)$$

yielding the equations for H and Ξ:

$$(1 - \eta^2) \frac{d^2 H}{d\eta^2} - 2\eta \frac{dH}{d\eta} + p(p+1)H = 0, \qquad (18.10)$$

$$(1 + \xi^2) \frac{d^2 \Xi}{d\xi^2} + 2\xi \frac{d\Xi}{d\xi} - p(p+1)\Xi = 0, \qquad (18.11)$$

where the separation constant has been written as $p(p + 1)$. Boundary conditions are

$$\left.\begin{array}{l} H = 0 \quad \text{at} \quad \eta = 0, H \text{ is well behaved } \text{ at } \eta = 1. \\ \Xi \to 0 \quad \text{as} \quad \xi \to \infty. \quad \Xi = 1 \quad \text{at} \quad \xi = 0. \end{array}\right\} \qquad (18.12)$$

The boundary condition for Ξ at $\xi = 0$ has been added in the anticipation of the possibility of superposing solutions to satisfy condition 18.8c.

Equation 18.10 is Legendre's equation, and we select the series solution 8.17 so as to satisfy the condition that $H = 0$ at $\eta = 0$. The series must terminate in order that H be well behaved at $\eta = 1$. This restricts p to odd integers

$$p = 2n + 1, n = 0, 1, 2, ..., \qquad (18.13)$$

and thus $H(\eta)$ is the Legendre polynomial

$$H(\eta) = P_{2n+1}(\eta). \qquad (18.14)$$

We now superpose solutions of this type and express the solution as

$$\Phi_2 = \sum_{n=0}^{\infty} C_n P_{2n+1}(\eta) \Xi_{2n+1}(\xi). \qquad (18.15)$$

We satisfy condition 18.8c by writing

$$-r_0 E_a \eta = \sum_{n=0}^{\infty} C_n P_{2n+1}(\eta) \Xi'_{2n+1}(0). \qquad (18.16)$$

Since the Legendre polynomials P_{2n+1} are orthogonal over the interval from $\eta = 0$ to $\eta = 1$, this allows us to evaluate the coefficients C_n:

$$C_n \Xi'_{2n+1}(0) \int_0^1 [P_{2n+1}(\eta)]^2 \, d\eta = -r_0 E_a \int_0^1 \eta P_{2n+1}(\eta) \, d\eta. \quad (18.17)$$

But $\eta = P_1(\eta)$ and is orthogonal to the other Legendre polynomials. Hence

$$C_0 \Xi'_1(0) = -r_0 E_a, \; C_n = 0, \, n = 1, 2, \ldots . \quad (18.18)$$

Now we solve for $\Xi_1(\xi)$ from the equation

$$(1 + \xi^2)\frac{d^2 \Xi_1}{d\xi^2} + 2\xi \frac{d\Xi_1}{d\xi} - 2\Xi_1 = 0 \quad (18.19)$$

and the boundary conditions

$$\Xi_1 = 1 \text{ at } \xi = 0, \quad \Xi_1 \to 0 \text{ as } \xi \to \infty. \quad (18.20)$$

Recognizing that $\Xi_1 = \xi$ is one solution to this homogeneous, linear, second-order equation, we work through the solution to obtain

$$\Xi_1 = A(1 + \xi\tan^{-1}\xi) + B\xi, \quad (18.21)$$

or

$$\Xi_1 = 1 + \xi\tan^{-1}\xi - \frac{\pi}{2}\xi, \quad (18.22)$$

so that

$$\Xi'_1(0) = -\pi/2. \quad (18.23)$$

Hence, the final solution for Φ_2 is

$$\Phi_2 = r_0 E_a \left[\frac{2}{\pi} - \xi\left(1 - \frac{2}{\pi}\tan^{-1}\xi\right)\right]\eta = \frac{2}{\pi}r_0 E_a(1 - \xi\operatorname{ctn}^{-1}\xi)\eta. \quad (18.24)$$

Thus, the solution for Φ is

$$\Phi = E_c z + E_a|z| + \frac{2}{\pi}r_0 E_a(1 - \xi\operatorname{ctn}^{-1}\xi)\eta = E_c z + \frac{2}{\pi}r_0 E_a(1 + \xi\tan^{-1}\xi)\eta.$$

$$(18.25)$$

Reference

1. Selçuk Ataman and D. N. Hanson. "Measurement of charged drops." *Industrial and Engineering Chemistry Fundamentals*, **8**, 833–836 (1969).

Chapter 19

Transient Temperature Distribution in a Slab

When the temperature in a solid varies with time, it obeys the equation

$$\frac{\partial T}{\partial t} = \alpha \nabla^2 T \, , \tag{19.1}$$

where α is the thermal diffusivity, assumed to be constant. We shall restrict ourselves here to temperature variations in the x-direction, with no variations in the y- or z-directions. Equation 19.1 then becomes the so-called heat equation

$$\frac{\partial T}{\partial t} = \alpha \frac{\partial^2 T}{\partial x^2} \, . \tag{19.2}$$

This equation is parabolic, and it is appropriate to have an initial condition at $t = 0$ as well as boundary conditions at two values of x.

By way of an illustrative example, we choose the conditions

$$\left. \begin{array}{l} T = 0 \quad \text{at} \quad x = 0 \\ \partial T/\partial x = 0 \quad \text{at} \quad x = L \\ T = f(x) \quad \text{at} \quad t = 0 \end{array} \right\}. \tag{19.3}$$

The Newman Lectures on Mathematics
John Newman and Vincent Battaglia
Copyright © 2018 Pan Stanford Publishing Pte. Ltd.
ISBN 978-981-4774-25-3 (Hardcover), 978-1-315-10885-8 (eBook)
www.panstanford.com

19.1 Solution by Separation of Variables. Series Solutions and Superposition

Assume a solution of the form

$$T(x, t) = P(t)X(x), \tag{19.4}$$

where P depends only on t and X depends only on x. Substitution into Eq. 19.2 gives

$$X\frac{dP}{dt} = \alpha P\frac{d^2 X}{dx^2}, \tag{19.5}$$

or after division by αPX:

$$\frac{1}{\alpha P}\frac{dP}{dt} = \frac{1}{X}\frac{d^2 X}{dx^2} = -\lambda^2, \tag{19.6}$$

where we have equated these terms to the separation constant $-\lambda^2$ since the left side cannot depend on x and the other term cannot depend on t.

The function $P(t)$, thus, satisfies the equation

$$dP/dt = -\lambda^2\alpha P \tag{19.7}$$

with the solution

$$P = e^{-\lambda^2\alpha t}, \tag{19.8}$$

where we have arbitrarily set the constant factor equal to 1. The function $X(x)$ satisfies the differential equation

$$d^2 X/dx^2 = -\lambda^2 X, \tag{19.9}$$

with boundary conditions derived from those in Eq. 19.3

$$dX/dx = 0 \text{ at } x = L \quad \text{and} \quad X = 0 \text{ at } x = 0. \tag{19.10}$$

The solution to this Sturm–Liouville system is

$$X = \sin\lambda x, \tag{19.11}$$

with λ restricted to the eigenvalues

$$\lambda_n L = \left(n - \frac{1}{2}\right)\pi, \quad n = 1, 2, \dots. \tag{19.12}$$

Here again the constant factor has been set equal to 1.

In this manner, we obtain a sequence of solutions corresponding to the eigenvalues of Eq. 19.12. A more general solution can be obtained by superposing these results. Thus, we express the temperature distribution as

$$T(x,t) = \sum_{n=1}^{\infty} C_n e^{-\lambda_n^2 \alpha t} \sin \lambda_n x .$$ (19.13)

The coefficients C_n are determined so that the initial condition is satisfied:

$$f(x) = \sum_{n=1}^{\infty} C_n \sin \lambda_n x .$$ (19.14)

Since the eigenfunctions are orthogonal, the coefficient C_k can be determined by multiplying by $\sin \lambda_k x$ and integrating from 0 to L:

$$\int_0^L f(x) \sin(\lambda_k x) dx = \sum_{n=1}^{\infty} C_n \int_0^L \sin(\lambda_n x) \sin(\lambda_k x) dx = C_k \int_0^L \sin^2(\lambda_k x) dx .$$

(19.15)

Hence

$$C_n = \frac{2}{L} \int_0^L f(x) \sin(\lambda_n x) dx .$$ (19.16)

We now take up the special case where $f(x) = T_0$, a constant. Then

$$C_n = \frac{2T_0}{\lambda_n L} .$$ (19.17)

The heat flux at the left boundary is of some interest and is given by

$$q_x(0,t) = -k \frac{\partial T}{\partial x}\bigg|_{x=0} = -\frac{2kT_0}{L} \sum_{n=1}^{\infty} e^{-\lambda_n^2 \alpha t} .$$ (19.18)

This is shown in Fig. 19.1, multiplied by $-\sqrt{\pi \alpha t}/kT_0$, as a function of $\alpha t/L^2$.

19.2 Similarity Solution for Short Times or Thick Slabs

The heat flux is infinite for $t = 0$ for the problem when $f(x) = T_0$. To obtain a concrete idea of this behavior, we can develop an asymptotic solution valid for small values of t. Then, the solution should be independent of the parameter L. Thus, at short times, T depends only on T and α as well as the independent variables x and t. For dimensional consistency, the solution must then be of the form

$$T = T_0 g(\eta) \text{ where } \eta = x/2\sqrt{\alpha t} . \qquad (19.19)$$

Thus, the temperature profiles at different times are similar in the sense that they can be made to coincide when expressed in terms of the variable η, a combination of the original independent variables.

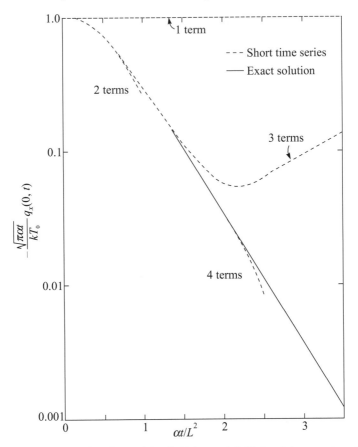

Figure 19.1 Heat flux for a slab of thickness L initially at temperature T_0. Temperature at the left boundary is 0, and the right boundary is insulated. Several terms of the short time solution are also shown.

A further consequence of Eq. 19.19 is that the function g must satisfy an ordinary differential equation with η as an independent variable. Substitution of Eq. 19.19 into Eq. 19.2 yields

$$\frac{d^2 g}{d\eta^2} + 2\eta \frac{dg}{d\eta} = 0, \tag{19.20}$$

and the boundary and initial conditions transform to

$$g = 0 \text{ at } \eta = 0 \quad \text{and} \quad g = 1 \text{ at } \eta = \infty. \tag{19.21}$$

The solution to Eq. 19.20 subject to conditions 19.21 is

$$g = \text{erf}(\eta). \tag{19.22}$$

Hence

$$T = T_0 \text{ erf}\left(x/2\sqrt{\alpha t}\right) \text{ for small } t. \tag{19.23}$$

From this result, we obtain the heat flux at the left boundary:

$$q_x(0,t) = -k\frac{\partial T}{\partial x}\bigg|_{x=0} = \frac{-kT_0}{2\sqrt{\alpha t}}\frac{dg}{d\eta}\bigg|_{\eta=0} = -\frac{kT_0}{\sqrt{\pi\alpha t}} \text{ for small } t.$$

$$\tag{19.24}$$

This shows that the heat flux is indeed infinite at $t = 0$, but it goes to infinity in such a way that the total heat flow from $t = 0$ to $t = t$ is finite. Figure 19.1 is plotted in such a way that the agreement of Eq. 19.18 with Eq. 19.24 at short times is evident.

19.3 Solution by the Method of Laplace Transforms

Laplace transforms are also useful for linear partial differential equations with constant coefficients where enough initial conditions are given. Here we use the Laplace transform with respect to t. Transformation of Eq. 19.2 gives

$$s\overline{T} - f(x) = \alpha\frac{\partial^2 \overline{T}}{\partial x^2}. \tag{19.25}$$

Here the transform $\overline{T}$ is a function of both s and x. The transformed Eq. 19.25 is effectively an ordinary differential equation in x, with the solution

$$\overline{T} = A \sinh\left(\sqrt{\frac{s}{\alpha}}x\right) + B \cosh\left(\sqrt{\frac{s}{\alpha}}x\right)$$

$$-\frac{1}{\sqrt{s\alpha}}\int_0^x \sinh\left(\sqrt{\frac{s}{\alpha}}(x-x')\right)f(x')dx'. \tag{19.26}$$

The transformed boundary conditions for Eq. 19.25 are

$$\overline{T} = 0 \quad \text{at} \quad x = 0 \quad \text{and} \quad \partial \overline{T}/\partial x = 0 \quad \text{at} \quad x = L. \qquad (19.27)$$

Hence, we set

$$B = 0$$

$$A = \frac{1}{\sqrt{\alpha s}\,\cosh\left(\sqrt{\dfrac{s}{\alpha}}\,L\right)} \left.\int_0^L \cosh\left(\sqrt{\dfrac{s}{\alpha}}(L - x')\right) f(x')dx'\right\}. \qquad (19.28)$$

For the special case where $f(x) = T_0$, a constant, we have

$$\overline{T} = \frac{T_0}{s}\tanh\left(\sqrt{\frac{s}{\alpha}}L\right)\sinh\left(\sqrt{\frac{s}{\alpha}}x\right) - \frac{T_0}{s}\cosh\left(\sqrt{\frac{s}{\alpha}}x\right) + \frac{T_0}{s}.$$

$$(19.29)$$

A useful feature of the Laplace transform method is that we can get the heat flux at the left boundary without inverting Eq. 19.29. Instead we obtain the transform of the temperature gradient at the left boundary:

$$\left.\frac{\partial \overline{T}}{\partial x}\right|_{x=0} = \frac{T_0}{\sqrt{\alpha s}}\tanh\left(\sqrt{\frac{s}{\alpha}}L\right). \qquad (19.30)$$

Although we may not find the inverse transform of Eq. 19.30 in the tables, we still can obtain useful results. Recall that the behavior of the Laplace transform for large values of s is related to the behavior of the time function for small values of t. Consequently, we expand Eq. 19.30 for large values of s:

$$\left.\frac{\partial \overline{T}}{\partial x}\right|_{x=0} = \frac{T_0}{\sqrt{\alpha s}}\frac{1 - e^{-2\sqrt{\frac{s}{\alpha}}L}}{1 + e^{-2\sqrt{\frac{s}{\alpha}}L}}$$

$$= \frac{T_0}{\sqrt{\alpha s}}\left(1 - e^{-2\sqrt{\frac{s}{\alpha}}L}\right)\left(1 - e^{-2\sqrt{\frac{s}{\alpha}}L} + e^{-4\sqrt{\frac{s}{\alpha}}L} - e^{-6\sqrt{\frac{s}{\alpha}}L} + \ldots\right)$$

$$= \frac{T_0}{\sqrt{\alpha s}}\left(1 - 2e^{-2\sqrt{\frac{s}{\alpha}}L} + 2e^{-4\sqrt{\frac{s}{\alpha}}L} - 2e^{-6\sqrt{\frac{s}{\alpha}}L} + \ldots\right).$$

$$(19.31)$$

Inversion, term by term, gives

$$q_x(0,t) = -k\frac{\partial T}{\partial x}\bigg|_{x=0}$$

$$= -\frac{kT_0}{\sqrt{\pi\alpha t}}\left[1 - 2e^{-L^2/\alpha t} + 2e^{-4L^2/\alpha t} - 2e^{-9L^2/\alpha t} + \ldots\right]. \quad (19.32)$$

The first term is in agreement with the short-time solution expressed in Eq. 19.24. The other terms give successively better agreement with the exact solution, as shown in Fig. 19.1.

We can obtain another series solution for the heat flux if we invert Eq. 19.30 by the method of residues (see Chapter 20, which should be studied before the remainder of the present chapter). This will be the same series as was obtained by the method of separation of variables, Eq. 19.18. The only singular points of the transform in Eq. 19.30 are at the roots of the denominator, $\cosh\left(\sqrt{\dfrac{s}{\alpha}}L\right)$. The function is analytic at $s = 0$.

The roots of $\cosh\left(\sqrt{\dfrac{s}{\alpha}}L\right)$ are given by

$$\sqrt{\frac{s}{\alpha}}L = i\left(n - \frac{1}{2}\right)\pi$$

or

$$s_n = -\left(n - \frac{1}{2}\right)^2\pi^2\alpha/L^2 = -\lambda_n^2\alpha, \quad n = 1, 2, \ldots . \quad (19.33)$$

The residue of $e^{st}\dfrac{T_0}{\sqrt{\alpha s}}\tanh\left(\sqrt{\dfrac{s}{\alpha}}L\right)$ at a singular point can be evaluated as follows, since these singular points are simple poles:

$$\rho_n = \lim_{s \to s_n}(s - s_n)e^{st}\frac{T_0}{\sqrt{\alpha s}}\tanh\left(\sqrt{\frac{s}{\alpha}}L\right)$$

$$= \frac{e^{s_n t}T_0}{\sqrt{\alpha s_n}}\sinh\left(\sqrt{\frac{s_n}{\alpha}}L\right)\lim_{s \to s_n}\frac{s - s_n}{\cosh\left(\sqrt{\dfrac{s}{\alpha}}L\right)}$$

$$= \frac{e^{s_n t} T_0}{\sqrt{\alpha s_n}} \sinh\left(\sqrt{\frac{s_n}{\alpha}}L\right) \frac{1}{\dfrac{1}{2}\dfrac{L}{\sqrt{\alpha s_n}} \sinh\left(\sqrt{\dfrac{s_n}{\alpha}}L\right)}$$

$$= \frac{2T_0}{L} e^{s_n t} = \frac{2T_0}{L} e^{-\lambda_n^2 \alpha t}. \tag{19.34}$$

The inversion integral for a Laplace transform $\overline{f}(s)$ is

$$\mathcal{L}^{-1}\left\{\overline{f}(s)\right\} = \frac{1}{2\pi i} \lim_{\beta \to \infty} \int_{\gamma - i\beta}^{\gamma + i\beta} e^{st} \overline{f}(s) ds, \tag{19.35}$$

where γ is large enough that all the singular points of $\overline{f}(s)$ lie to the left of the path of integration. When all the singular points of $\overline{f}(s)$ are isolated from each other, as in the present case, this integral can be replaced by the sum of integrals around each of the singular points:

$$\mathcal{L}^{-1}\left\{\overline{f}(s)\right\} = \sum_n \rho_n(t), \tag{19.36}$$

where ρ_n is the residue of the integrand at s_n:

$$\rho_n = \frac{1}{2\pi i} \oint_{s_n} e^{st} \overline{f}(s) ds. \tag{19.37}$$

In the case of a singular point that is a simple pole, the residue can be evaluated in the manner of Eq. 19.34. Hence, the inverse of Eq. 19.30, obtained by the method of residues, can be written as

$$\left.\frac{\partial T}{\partial x}\right|_{x=0} = \sum_n \rho_n = \frac{2T_0}{L} \sum_{n=1}^{\infty} e^{-\lambda_n^2 \alpha t}. \tag{19.38}$$

This gives a series for the heat flux at the left boundary, which is the same as Eq. 19.18.

In this way, the Laplace transform method can be made to yield two series solutions, one (Eq. 19.32) that converges rapidly at short times and one (Eq. 19.38) that converges rapidly at large times.

Chapter 20

Inversion of Laplace Transforms by the Method of Residues

The inverse of a Laplace transform $\overline{f}(s)$ can be expressed in terms of the inversion integral:

$$f(t) = \mathscr{L}^{-1}\{\overline{f}(s)\} = \frac{1}{2\pi i} \lim_{\beta \to \infty} \int_{\gamma-i\beta}^{\gamma+i\beta} e^{st}\,\overline{f}(s)ds. \qquad (20.1)$$

This involves a complex integration in the complex plane, and γ must be large enough that $\overline{f}(s)$ is analytic (i.e., can be expanded in a Taylor series) in the plane to the right of the contour of integration (i.e., $\overline{f}(s)$ must be analytic at all points s such that $Re\{s\} \geq \gamma$). In other words, all the singular behavior of $\overline{f}(s)$ must lie to the left of the contour of integration. The inversion integral is analogous to the inverse of a Fourier transform.

Integrations in the complex plane have the useful property that the value of the integral from one point to another is independent of the path of integration if the integrand is analytic on two such paths and in the region enclosed by them. Thus, the above integral is independent of the value of γ as long as γ is large enough that the integrand is analytic to the right of the path of integration.

If $\overline{f}(s)$ is not analytic only at isolated "singular" points, then the path of integration of Eq. 20.1 can be extended to include a path

The Newman Lectures on Mathematics
John Newman and Vincent Battaglia
Copyright © 2018 Pan Stanford Publishing Pte. Ltd.
ISBN 978-981-4774-25-3 (Hardcover), 978-1-315-10885-8 (eBook)
www.panstanford.com

around the entire left half of the complex plane, as shown in Fig. 20.1. In these cases, the integrand behaves so that the contribution along this extended path of integration is zero (as $\beta \to \infty$). This then gives a closed path of integration. Furthermore, since the integrand is analytic except at the isolated singular points and since in an analytic region, the integral is independent of the path, the contour shown in Fig. 20.1a can be deformed so that the path goes around each of the singular points. This leads to the representation of the inversion integral as a sum of "residues"

$$f(t) = \mathcal{L}^{-1}\{\overline{f}(s)\} = \sum_{n=1}^{\infty} \rho_n(\overline{f}(s)e^{st}; s_n), \tag{20.2}$$

where the residue of the function $e^{st}\overline{f}(s)$ at an isolated singularity s_n is given by the integral around that point, divided by $2\pi i$:

$$\rho_n(e^{st}\overline{f}(s); s_n) = \frac{1}{2\pi i}\oint_{s_n} e^{st}\overline{f}(s)ds = \rho_n(t). \tag{20.3}$$

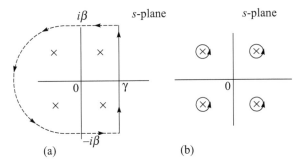

(a)   (b)

Figure 20.1 (a) For a case with isolated singularities, the basic path of the inversion integral (shown solid) can be extended by the dashed contour to form a closed contour. Representative, isolated singular points are shown by x. (b). Since the integrand is analytic accept at the isolated singular points, the contour integration shown in (a) is equivalent to the sum of the integrations around each singular point, as shown in (b).

This method cannot be applied if the integrand has singular points that are not isolated. The most common occurrence of this behavior is with branch points. For example, $\sqrt{s}$ and $\ln s$ are not single valued unless we consider only one branch of the functions. If we write $s = re^{i\theta}$, then

$$\sqrt{s} = \sqrt{r}e^{i\theta/2}, -\pi \le \theta < \pi \qquad (20.4)$$

$$\ln s = \ln r + i\theta, -\pi \le \theta < \pi, \qquad (20.5)$$

where we have arbitrarily selected one branch so that the functions are single valued. However, these functions are then discontinuous, and hence not analytic, along the negative real axis. If the integrand in Eq. 20.1 has singular points that are not isolated, then the method of residues cannot be applied in a straightforward manner.

For the fortunate cases of isolated singularities, we are then faced with the problem of evaluating the residues. Near any isolated singularity, $e^{st}\overline{f}(s)$ can always be expanded in a Laurent series of the form

$$e^{st}\overline{f}(s) = \sum_{k=-\infty}^{\infty} a_k(s - s_n)^k. \qquad (20.6)$$

Since

$$\oint (s - s_n)^k ds = 0 \text{ if } k \text{ is an integer not equal to } -1, \qquad (20.7)$$

$$\oint \frac{ds}{s - s_n} = 2\pi i, \qquad (20.8)$$

it follows that the residue at s_n is given by $\rho_n = a_{-1}$, the coefficient of $1/(s - s_n)$ in the Laurent series of $e^{st}\overline{f}(s)$ valid in the neighborhood of s_n. This gives us one way to obtain the residue at s_n, namely, expand $e^{st}\overline{f}(s)$ in a Laurent series valid near s_n and select the coefficient of $1/(s - s_n)$. This may be easier than carrying out the integration in Eq. 20.3.

In the special, but frequently encountered, case of a "simple pole," for which the Laurent series is

$$e^{st}\overline{f}(s) = \sum_{k=-1}^{\infty} a_k(s - s_n)^k, \qquad (20.9)$$

the residue can be evaluated in the relatively simple manner:

$$\rho_n = \lim_{s \to s_n}(s - s_n)e^{st}\overline{f}(s) = e^{s_n t}\lim_{s \to s_n}(s - s_n)\overline{f}(s). \qquad (20.10)$$

These comments should be helpful in solving problems involving the inversion of Laplace transforms.

Problem

20.1 Equation 19.29 represents the Laplace transform of the temperature for transient heat conduction in a slab with the initial condition $T = f(x) = T_0$ at $t = 0$ and the boundary conditions $T = 0$ at $x = 0$ and $\partial T/\partial x = 0$ at $x = L$. We used this to obtain two series expressions for the heat flux at the left boundary. In the same manner, invert the transform 19.29 to get the temperature distribution as follows:

(a) Obtain an expansion for T valid at short times by expanding the transform for large values of s and then inverting term by term. Compare the first term with the similarity solution obtained in Chapter 19. Note the inverse transform

$$\mathscr{L}^{-1}\left\{\frac{1}{s}e^{-k\sqrt{s}}\right\} = \operatorname{erfc}\left(\frac{k}{2\sqrt{t}}\right) = \frac{2}{\sqrt{\pi}}\int_{k/2\sqrt{t}}^{\infty} e^{-x^2}dx.$$

(b) Use the method of residues to obtain an expansion, useful for larger times, which coincides with that obtained in Chapter 19 by the method of separation of variables.

Chapter 21

Similarity Transformations

Similarity transformations are really quite useful because they allow partial differential equations, even nonlinear ones, to be reduced to ordinary differential equations. However, all similarity transformations cannot be discovered by the methods of dimensional analysis, and it is worthwhile to consider how to find a similarity transformation.

Let us look again at the problem

$$\frac{\partial T}{\partial t} = \alpha \frac{\partial^2 T}{\partial x^2} \,, \tag{21.1}$$

$$T = T_0 \text{ at } t = 0, \quad T = 0 \text{ at } x = 0, \quad T \to T_0 \text{ as } x \to \infty. \tag{21.2}$$

We can seek a similarity transformation by assuming tentatively that

$$T = T(\eta) \quad \text{where} \quad \eta = x/h(t). \tag{21.3}$$

The similarity variable η introduced here can be seen to be fairly general. However, a more general form for the function would be

$$T = f(x)g(\eta), \tag{21.4}$$

because, as we saw in Problem 15.1, our transformation may look like

$$T = T_0(1 - x/L)g(\eta). \tag{21.5}$$

But let us return to the form in Eq. 21.3. Substitution into Eq. 21.1 gives

The Newman Lectures on Mathematics
John Newman and Vincent Battaglia
Copyright © 2018 Pan Stanford Publishing Pte. Ltd.
ISBN 978-981-4774-25-3 (Hardcover), 978-1-315-10885-8 (eBook)
www.panstanford.com

$$\frac{dT}{d\eta}\left(-\frac{x}{h^2}\frac{dh}{dt}\right) = \alpha\frac{d^2T}{d\eta^2}\frac{1}{h^2} \tag{21.6}$$

or

$$\eta\frac{dT}{d\eta}\left(\frac{h}{\alpha}\frac{dh}{dt}\right) + \frac{d^2T}{d\eta^2} = 0. \tag{21.7}$$

If the similarity transformation is to work, the variables t and x cannot appear separately in Eq. 21.7, only in the combination $\eta = x/h(t)$. This can be accomplished in the present case by setting the term in parentheses equal to a constant, where we choose to use 2 for simplification in later expressions. Thus, we obtain two ordinary differential equations:

$$\frac{d^2T}{d\eta^2} + 2\eta\frac{dT}{d\eta} = 0, \tag{21.8}$$

$$\frac{h}{\alpha}\frac{dh}{dt} = \frac{1}{2\alpha}\frac{dh^2}{dt} = 2, \tag{21.9}$$

with the solutions

$$T = A\int_0^{\eta} e^{-x^2}dx + B, \tag{21.10}$$

$$h^2 = 4\alpha t + C. \tag{21.11}$$

We should set $C = 0$ because we must also collapse two boundary conditions into one:

$$T \to T_0 \text{ as } \eta \to \infty. \tag{21.12}$$

Then we have

$$\eta = x/2\sqrt{\alpha t} \tag{21.13}$$

and the other boundary condition for Eq. 21.8 or 21.10 is

$$T = 0 \quad \text{at} \quad \eta = 0. \tag{21.14}$$

To satisfy the boundary conditions, Eq. 21.10 becomes

$$T = T_0\frac{2}{\sqrt{\pi}}\int_0^{\eta} e^{-x^2}dx = T_0\text{erf}\left(x/2\sqrt{\alpha t}\right). \tag{21.15}$$

Now let us consider mass transfer to a spherical drop whose radius increases with time in an arbitrary manner. The governing equation is Eq. 13.26 of convective diffusion, which reads in spherical coordinates

$$\frac{\partial c}{\partial t} + v_r \frac{\partial c}{\partial r} = \frac{D}{r^2}\frac{\partial}{\partial r}\left(r^2\frac{\partial c}{\partial r}\right) = D\frac{\partial^2 c}{\partial r^2} + \frac{2D}{r}\frac{\partial c}{\partial r}, \qquad (21.16)$$

for spherical symmetry. For boundary conditions, we take

$$c = 0 \text{ at } r = r_0(t) \quad \text{and} \quad c \to c_\infty \text{ as } r \to \infty, \qquad (21.17)$$

where r_0 is the radius of the drop and depends on time. We shall, at the outset, assume that the diffusion layer, in which the concentration deviates appreciably from the bulk value c_∞, is thin compared to the radius of the drop and that consequently the second term on the right in Eq. 21.16 can be neglected.

The radial velocity is determined by the rate of growth of the drop:

$$4\pi r^2 v_r = 4\pi r_0^2 \, dr_0/dt \qquad (21.18)$$

or

$$v_r = \frac{r_0^2}{r^2}\frac{dr_0}{dt}. \qquad (21.19)$$

This is equivalent to the use of the continuity Eq. 13.21 for an incompressible fluid.

Now let us change to the variable

$$y = r - r_0(t), \qquad (21.20)$$

the normal distance from the surface of the drop. The coordinate transformation from r, t to y, t gives

$$\left(\frac{\partial c}{\partial t}\right)_r = \left(\frac{\partial c}{\partial t}\right)_y\left(\frac{\partial t}{\partial t}\right)_r + \left(\frac{\partial c}{\partial y}\right)_t\left(\frac{\partial y}{\partial t}\right)_r = \frac{\partial c}{\partial t} - \frac{dr_0}{dt}\frac{\partial c}{\partial y} \qquad (21.21)$$

$$\left(\frac{\partial c}{\partial r}\right)_t = \left(\frac{\partial c}{\partial t}\right)_y\left(\frac{\partial t}{\partial r}\right)_t + \left(\frac{\partial c}{\partial y}\right)_t\left(\frac{\partial y}{\partial r}\right)_t = \frac{\partial c}{\partial y} \qquad (21.22)$$

$$\left(\frac{\partial^2 c}{\partial r^2}\right)_t = \frac{\partial^2 c}{\partial y^2}. \qquad (21.23)$$

Hence, Eq. 21.16 becomes

$$\frac{\partial c}{\partial t} - \frac{dr_0}{dt}\frac{\partial c}{\partial y} + \frac{r_0^2}{r^2}\frac{dr_0}{dt}\frac{\partial c}{\partial y} = D\frac{\partial^2 c}{\partial y^2}, \tag{21.24}$$

or

$$\frac{\partial c}{\partial t} - \left(1 - \frac{r_0^2}{r^2}\right)\frac{dr_0}{dt}\frac{\partial c}{\partial y} = D\frac{\partial^2 c}{\partial y^2}, \tag{21.25}$$

or

$$\frac{\partial c}{\partial t} - \frac{(r - r_0)(r + r_0)}{r^2}\frac{dr_0}{dt}\frac{\partial c}{\partial y} = D\frac{\partial^2 c}{\partial y^2}. \tag{21.26}$$

Again on the basis of the thinness of the diffusion layer compared to the radius of the drop, we approximate $r + r_0$ by $2r_0$ and approximate r^2 by r_0^2, and Eq. 21.26 becomes

$$\frac{\partial c}{\partial t} - \frac{2y}{r_0}\frac{dr_0}{dt}\frac{\partial c}{\partial y} = D\frac{\partial^2 c}{\partial y^2}. \tag{21.27}$$

Now seek a similarity solution of the form

$$c = c(\eta) \quad \text{where} \quad \eta = y/h(t). \tag{21.28}$$

Substitution into Eq. 21.27 gives

$$\frac{dc}{d\eta}\left(-\frac{y}{h^2}\frac{dh}{dt}\right) - \frac{2y}{r_0}\frac{dr_0}{dt}\frac{dc}{d\eta}\frac{1}{h} = D\frac{d^2 c}{d\eta^2}\frac{1}{h^2} \tag{21.29}$$

or

$$\frac{d^2 c}{d\eta^2} + \frac{\eta}{D}\frac{dc}{d\eta}\left(h\frac{dh}{dt} + 2\frac{h^2}{r_0}\frac{dr_0}{dt}\right) = 0. \tag{21.30}$$

As in the case of Eq. 21.7, the similarity transformation is successful only if the variables t and y appear in Eq. 21.30 only in the combination $\eta = y/h(t)$. Consequently, we set the term in parentheses equal to a constant, $2D$, and thereby generate two ordinary differential equations

$$\frac{d^2 c}{d\eta^2} + 2\eta\frac{dc}{d\eta} = 0 \tag{21.31}$$

$$h\frac{dh}{dt} + 2\frac{h^2}{r_0}\frac{dr_0}{dt} = 2D. \tag{21.32}$$

Equation 21.32 is a linear, first-order differential equation for h^2,

$$\frac{dh^2}{dt} + 4\frac{h^2}{r_0}\frac{dr_0}{dt} = 4D, \tag{21.33}$$

with the solution (see Chapter 2)

$$h^2 r_0^4 = \int_0^t 4Dr_0^4 dt + C. \tag{21.34}$$

If we impose the initial condition

$$c = c_\infty \text{ at } t = 0, \tag{21.35}$$

then we should set the integration constant C equal to zero so that this initial condition and the boundary condition at $r = \infty$ collapse into a single boundary condition

$$c \to c_\infty \text{ as } \eta \to \infty, \tag{21.36}$$

and the similarity variable becomes

$$\eta = yr_0^2 \bigg/ \left[4D\int_0^t r_0^4 dt \right]^{1/2}. \tag{21.37}$$

The boundary condition at $r = r_0$ becomes

$$c = 0 \text{ at } \eta = 0, \tag{21.38}$$

and the solution to Eq. 21.31 becomes

$$c = c_\infty \frac{2}{\sqrt{\pi}} \int_0^\eta e^{-x^2} dx = c_\infty \text{erf}(\eta), \tag{21.39}$$

where η is given by Eq. 21.37.

This completes our solution to the problem of mass transfer to a spherical drop whose radius increases with time in an arbitrary manner. The similarity transformation 21.37 could not be arrived at by dimensional arguments alone, and it is helpful to have an orderly procedure for searching for such transformations. In the course of obtaining the solution, we made several approximations; the term $(2D/r)\partial c/\partial r$ was neglected in Eq. 21.16 and y was neglected compared to r_0 in going from Eq. 21.26 to Eq. 21.27. It should be possible to justify these approximations *a posteriori* and to use the solution obtained above as a basis for obtaining higher-order approximations. This is carried out in Ref. [1] for the case where the volume of the drop increases linearly with time.

Sedov gives a variety of similarity solutions to partial differential equations [2].

References

1. John Newman. "The Koutecký correction to the Ilkovič equation." *Journal of Electroanalytical Chemistry and Interfacial Electrochemistry*, **15**, 309–312 (1967).
2. L. I. Sedov. *Similarity and Dimensional Methods in Mechanics.* New York: Academic Press (1959).

Problems

21.1 Show that the total rate of mass transfer to a drop growing at a constant volumetric rate is

$$4\sqrt{\frac{7\pi D}{3}}\left(\frac{3Q}{4\pi}\right)^{2/3} c_{\infty} t^{1/6}$$

where Q is the volumetric rate of growth (cm^3/s). This is the result of Ilkovič for the rate of mass transfer to a drop growing at the tip of a capillary.

21.2 A gas bubble grows spherically in a supersaturated solution. Suppose that the convection is due entirely to the growth of the bubble, and show that the size of the bubble is given by

$$r_0 = 2\sqrt{\frac{3Dt}{\pi}\frac{RT}{p}\left(c_{\infty} - c_0\right)}$$

where t is the time since nucleation of the bubble, D is the diffusion coefficient of the gas dissolved in the solution, R is the gas constant, T is the absolute temperature, p is the pressure inside the bubble taken to be constant, c_{∞} is the concentration of the gas in the supersaturated bulk solution, and c_0 is the saturation concentration prevailing at the surface of the bubble.

Chapter 22

Superposition Integrals and Integral Equations

For heat transfer in a semi-infinite slab, we can take the governing equation to be

$$\frac{\partial T}{\partial t} = \alpha \frac{\partial^2 T}{\partial x^2} \tag{22.1}$$

with the conditions

$$T = 0 \quad \text{at} \quad t = 0, \tag{22.2}$$

$$T = g(t) \quad \text{at} \quad x = 0, \tag{22.3}$$

and

$$T \rightarrow 0 \quad \text{as} \quad x \rightarrow \infty. \tag{22.4}$$

The first and the last conditions have been restated so that T becomes zero.

We can first develop a solution for a step change at the boundary, having $T = T_0$ at $x = 0$. The similarity solution is a slight modification of Eq. 19.23:

$$T = T_0 \operatorname{erfc}\left(\frac{x}{2\sqrt{\alpha t}}\right) = T_0 \frac{2}{\sqrt{\pi}} \int\limits_{x/2\sqrt{\alpha t}}^{\infty} e^{-\eta^2} d\eta. \tag{22.5}$$

Now, in preparation for the method of superposition integrals, solve the above problem with the boundary condition $T = g(t)$ at $x = 0$ by the method of Laplace transforms, with which you are very

The Newman Lectures on Mathematics
John Newman and Vincent Battaglia
Copyright © 2018 Pan Stanford Publishing Pte. Ltd.
ISBN 978-981-4774-25-3 (Hardcover), 978-1-315-10885-8 (eBook)
www.panstanford.com

familiar by now. Take the Laplace transform of the differential equation with respect to t:

$$s\overline{T}(x,s) - 0 = \alpha \frac{\partial^2 \overline{T}}{\partial x^2}. \tag{22.6}$$

This looks like an ordinary differential equation with constant coefficients. The solution is

$$\overline{T} = A(s)e^{-x\sqrt{s/\alpha}} + B(s)e^{x\sqrt{s/\alpha}}. \tag{22.7}$$

We set $B(s) = 0$ so that we can satisfy the boundary condition

$$\overline{T}(x, s) \to 0 \quad \text{as} \quad x \to \infty. \tag{22.8}$$

To get $A(s)$, take the Laplace transform of the boundary condition at $x = 0$:

$$\overline{T}(x, s) = G(s) \quad \text{at} \quad x = 0. \tag{22.9}$$

Comparison with the solution obtained from the differential equation shows that $A(s) = G(s)$, and hence

$$\overline{T}(x, s) = G(s)e^{-x\sqrt{s/\alpha}}. \tag{22.10}$$

We use the convolution integral to effect the inversion of the Laplace transform. We look up in the tables that

$$\mathcal{L}^{-1}\left\{\frac{1}{s}e^{-x\sqrt{s/\alpha}}\right\} = \text{erfc}\left(\frac{x}{2\sqrt{\alpha t}}\right), \tag{22.11}$$

and we note that this is what we obtained earlier if we set $T_0 = 1$. Let us call this solution

$$T_1(x,t) = \text{erfc}\left(\frac{x}{2\sqrt{\alpha t}}\right), \tag{22.12}$$

that is, the solution to the problem for a unit step of the temperature at the boundary (at $x = 0$) with the unit step made at $t = 0$ and applied thereafter. Therefore, $\overline{T}(x, s)$ is seen to be the product of two Laplace transforms, $sG(s)$ and $(1/s)\exp(-x\sqrt{s/\alpha})$, which we rewrite as

$$\overline{T}(x,s) = [sG(s) - g(0+)]\frac{e^{-x\sqrt{s/\alpha}}}{s} + g(0+)\frac{e^{-x\sqrt{s/\alpha}}}{s}, \tag{22.13}$$

so that it is clear that the inverse can be written as

$$T(x,t) = \int_0^t \frac{dg}{dt}\bigg|_{t=t'} T_1(x,\ t-t')dt' + g(0+)T_1(x,t). \quad (22.14)$$

This result can be regarded as a superposition integral; the function $T_1(x, t - t')$ shows the influence at x and t of a modification of the boundary temperature, a change dg/dt, made at $t = t'$.

22.1 Duhamel's Theorem

One way to think about a superposition integral like in the preceding example is to divide the time domain for $g(t)$ into intervals and to approximate g by a constant within each interval (see Fig. 22.1). Because the differential equation and other boundary conditions are linear, one can superpose solutions corresponding to the jump in the g approximation from one interval to the next.

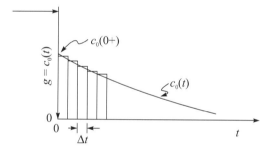

Figure 22.1 Approximation of the surface concentration $c_0(t)$ by a series of steps. For the first example, interpret $c_0(t)$ as $T_0(t)$.

Say that we have a jump Δg at a time t_i. We can define a function $\Delta g T_1(x,\ t - t_i)$ corresponding to the response to this jump. The responses to prior jumps continue, and the responses to jumps occurring at times greater than t have not yet been felt. Thus, an approximation to the function $T(x, t)$ is a sum over all the applicable responses to jumps.

Duhamel's theorem says, essentially, that the limit of this sum as we shorten the jump interval (and increase their number) is the superposition integral in Eq. 22.14. That example provides a concrete illustration of the principle of superposition and of superposition integrals.

The principle of superposition was introduced in the chapter on linear systems. It permits us to break down a problem into simpler parts, solve these separately, and add the results to get the solution to the original problem. Because linear systems are well studied, there are diverse problems to illustrate this.

22.2   Integral Equations

Suppose in the original problem that the boundary condition $T = g(t)$ at $x = 0$ is replaced by a more general, perhaps nonlinear, relationship between T and $\partial T / \partial x$ at $x = 0$. We can obtain an integral equation for the unknown boundary temperature by differentiating the basic solution 22.14:

$$\frac{\partial T}{\partial x} = \int_0^t \frac{dg}{dt}\bigg|_{t=t'} \frac{\partial T_1}{\partial x}\bigg|_{t=t-t'} dt' + g(0+)\frac{\partial T_1}{\partial x}\bigg|_t . \qquad (22.15)$$

For this example, use the explicit solution in Eq. 22.12, and evaluate the result at $x = 0$:

$$\frac{\partial T_1}{\partial x}\bigg|_{x=0} = \frac{-1}{\sqrt{\pi \alpha t}} . \qquad (22.16)$$

Consequently, Eq. 22.15 can be written as

$$\frac{\partial T}{\partial x}\bigg|_{x=0} = -\int_0^t \frac{dg}{dt}\bigg|_{t=t'} \frac{dt'}{\sqrt{\pi \alpha (t-t')}} - \frac{g(0+)}{\sqrt{\pi \alpha t}} . \qquad (22.17)$$

This equation thus provides a relationship between T and the normal derivative of T at the boundary. If there is another physical relationship between these two quantities, one has an integral equation to solve.

For simplicity of notation, let

$$h(t) = \frac{\partial T}{\partial x}\bigg|_{x=0} . \qquad (22.18)$$

The above equation becomes

$$h = -\int_0^t g'(t')\frac{dt'}{\sqrt{\pi \alpha (t-t')}} - \frac{g(0+)}{\sqrt{\pi \alpha t}} . \qquad (22.19)$$

Suppose that h depends on g (in addition possibly to t). Then the integral equation to solve for $g(t)$ can be written as

$$h(t,g) = -\int_0^t \frac{dg}{dt}\bigg|_{t'} \frac{dt'}{\sqrt{\pi\alpha(t-t')}} - \frac{g(0+)}{\sqrt{\pi\alpha t}}. \qquad (22.20)$$

Regard this as an equation for the unknown $g(t)$.

One example where h would depend on g would be radiant heat transfer at the surface. Another, more interesting example is where T represents the concentration c, α represents the diffusion coefficient $\mathcal{D}$, and the boundary condition represents a heterogeneous reaction that consumes the species represented by c (see Section 22.3).

The integral equation thus developed can be solved numerically (see Wagner [1] or Acrivos and Chambré [2] for a similar problem). Such a solution is simpler and inherently more accurate than solving the original partial differential equation numerically because $g(t)$ will represent fewer unknowns (after introducing some finite-difference form) and because careful attention can be devoted to any singular behavior of the integral equation. After $g(t)$ is obtained, the profiles $T(x, t)$ can be obtained from Eq. 22.14, where g is now known.

For some purposes, it may be useful to have an alternative integral equation that puts $g(t)$ on the left and expresses this as an integral over the normal derivative h. Again, the relationship between g and h would be substituted in to obtain an integral equation for g or h. You are encouraged in Problem 22.2 to carry this out for the above example, with the result

$$g(t) = -\sqrt{\alpha} \int_0^t h(t') \frac{dt'}{\sqrt{\pi(t-t')}}. \qquad (22.21)$$

Regarded as a superposition integral, the solution $-\sqrt{\alpha/\pi(t-t')}$ represents the value of $T(0, t)$ at the surface resulting from the flux density in the form of a Dirac delta function occurring at the surface at $t = t'$. This flux density represented by $h(t')$ is multiplied by this function and integrated over t'. Thus, Eq. 22.19 is a superposition integral for the effect of a step change in T at the surface (at $x = 0$) at $t = t'$ on the flux density h on the surface at t, while Eq. 22.21 is a superposition integral for the effect of a flux source at the surface at $t = t'$ on the value of T on the surface at t.

22.3   Catalytic Reaction at a Surface

Let us carry this out for a second-order heterogeneous reaction at the surface. The problem should then be restated in terms of mass transfer, with T becoming c and with α becoming $\mathcal{D}$. Thus, the boundary condition is taken to be

$$-\mathcal{D}\frac{\partial c}{\partial x}\bigg|_{x=0} = -kc_0^2 . \tag{22.22}$$

This is deliberately taken to be nonlinear so that you can see that the last boundary condition can be nonlinear, even though the partial differential equation and the other boundary and initial conditions should be linear. Now substitution into the superposition-integral expression gives

$$\frac{k}{\mathcal{D}}c_0^2 = -\frac{1}{\sqrt{\pi\mathcal{D}}}\int_0^t \frac{dc_0}{dt}\bigg|_{t=t'} \frac{dt'}{\sqrt{t-t'}} - \frac{c_0(0+)-c_\infty}{\sqrt{\pi\mathcal{D}t}} . \tag{22.23}$$

Here the unknown is $c_0(t)$; this appears both on the left and in the integrand. This gives an *integral equation* for $c_0(t)$.

How should we approach the problem numerically? First, let us get rid of the parameters c_∞, $\mathcal{D}$, and k. Let $\Theta = c_0/c_\infty$. The integral equation becomes

$$\frac{k}{\mathcal{D}}c_\infty\Theta^2 = -\frac{1}{\sqrt{\pi\mathcal{D}}}\int_0^t \frac{d\Theta}{dt}\bigg|_{t=t'} \frac{dt'}{\sqrt{t-t'}} - \frac{\Theta(0+)-1}{\sqrt{\pi\mathcal{D}t}} . \tag{22.24}$$

Next let $\tau = c_\infty^2 k^2 t/\mathcal{D}$. (You can verify that this is dimensionless.) The integral equation becomes

$$-\Theta^2 = \frac{1}{\sqrt{\pi}}\int_0^\tau \frac{d\Theta}{d\tau}\bigg|_{\tau=\tau'} \frac{d\tau'}{\sqrt{\tau-\tau'}} + \frac{\Theta(0+)-1}{\sqrt{\pi\tau}} . \tag{22.25}$$

This gets rid of the parameters and leaves us with a simple integral equation for the function Θ as a function of τ.

Next, one can argue that $\Theta(0+) = 1$ or

$$\Theta = 1 \quad \text{at} \quad \tau = 0. \tag{22.26}$$

In other words, the concentration does not change discontinuously at $\tau = 0$; if it did, this would require an infinite flux density, which would not be in agreement with the kinetics, which gives a finite rate.

The problem now reads

$$-\Theta^2 = \frac{1}{\sqrt{\pi}} \int_0^\tau \frac{d\Theta}{d\tau}\bigg|_{\tau=\tau'} \frac{d\tau'}{\sqrt{\tau-\tau'}}, \quad \text{with } \Theta = 1 \text{ at } \tau = 0. \quad (22.27)$$

What can we get out of the problem by inspection, before we solve it numerically? We can see that at short times $\Theta \approx 1$. We can say that the rate is kinetically limited. We can take Θ^2 itself as a dimensionless reaction rate. At long times, $\Theta \to 0$, and the rate becomes mass-transfer limited. At very long times, it will look like the change in Θ, $d\Theta/d\tau$, occurred near $\tau = 0$, and the above equation suggests that

$$-\Theta^2 \to \frac{1}{\sqrt{\pi}}(-1)\frac{1}{\sqrt{\tau}} \quad \text{or} \quad \Theta^2 \to \frac{1}{\sqrt{\pi\tau}}. \quad (22.28)$$

We can make a guess at the behavior, as shown in Fig. 22.2. This is shown quantitatively in Additional Notes (Fig. 22.5), after we have solved the problem numerically, but we can gain a very good idea of the behavior of the solution without actually carrying out the numerics.

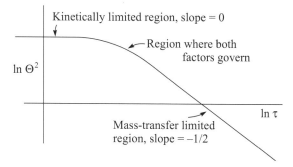

Figure 22.2 Sketch of the behavior for the reaction rate as a function of time for a second-order heterogeneous reaction at the wall of a semi-infinite stagnant region.

22.4 Superposition Integral from Laplace's Equation

Let us consider a cylindrical geometry where a line source of current at (x', y') leads to a solution of Laplace's equation as follows:

$$\Phi = \frac{I}{2\pi\kappa} \ln\left[(x - x')^2 + (y - y')^2\right], \qquad (22.29)$$

where I is the strength of the line current (in amperes per meter) and κ is the conductivity (S/m) of the medium. This represents a current flowing radially outward from the line at (x', y'), as shown in Fig. 22.3.

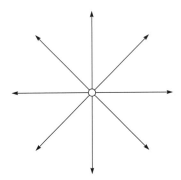

Figure 22.3   Current flow from a line source, a solution to Laplace's equation.

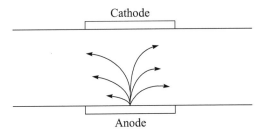

Cathode

Anode

Figure 22.4   A line source of current on the wall of a flow channel.

A variant on this is a line source between two insulating planes (see Fig. 22.4). Let us take the anode of our system to lie in the plane at $y = 0$, so that the solution to Laplace's equation for a line source at $(0, x')$ is

$$\Phi = \frac{I}{2\pi\kappa} \ln\left[\sinh^2\left(\frac{\pi(x - x')}{2h}\right) + \sin^2\left(\frac{\pi y}{2h}\right)\right], \qquad (22.30)$$

where h is the distance between the two planes and I is the current (per unit length) on the side that enters the flow channel. The reader may want to verify that $\partial\Phi/\partial y$ is zero at the plane at $y = h$ and also at the plane at $y = 0$ (except at the point x', 0). The current generated along the line source flows away from the line source between the two insulating planes, adopting a linear potential profile for x approaching either $+\infty$ or $-\infty$.

We can construct a superposition integral because Laplace's equation is linear. Let us do this all at once, for a cathode located in the plane at $y = h$ and the anode in the plane $y = 0$:

$$\Phi = \Phi^* - \frac{1}{2\pi\kappa} \int_0^L i_{cath}(x') \ln\left[\sinh^2\left(\frac{\pi(x - x')}{2h} \right) + \sin^2\left(\frac{\pi(y - h)}{2h} \right) \right] dx'$$

$$- \frac{1}{2\pi\kappa} \int_0^L i_{anode}(x') \ln\left[\sinh^2\left(\frac{\pi(x - x')}{2h} \right) + \sin^2\left(\frac{\pi y}{2h} \right) \right] dx' ,$$

$$(22.31)$$

where Φ^* is an additive constant representing the arbitrary zero of potential in electrolytic systems. The equation should be contemplated carefully. From the basic solutions used to construct the superposition integral, the current densities along the two electrodes are $i_{cath}(x')$ on the cathode at $y = h$ and $i_{anode}(x')$ on the anode at $y = 0$. The current from each line source can be regarded as flowing off to infinity between the two insulating planes. However, by requiring that the integrals of the current–density distributions be the same but of opposite sign on the two electrodes, this represents a system with current flowing from the anode to the cathode. In contrast to the previous examples, the upper limit of the integral is L instead of x. This reflects the fact that Laplace's equation is elliptic and all points can affect all other points, whereas the previous examples represented a parabolic equation where the future does not affect the past.

By introducing equations describing the dependence of i_{anode} and i_{cath} on the potential differences at each electrode interface, one can obtain integral equations for determining the current distribution along the two electrodes.

22.5 Wealth of Superposition Integrals

Since superposition is a general property of linear systems, many situations arise where superposition integrals are important. Let us list some of these situations:

1. *Coupled species, with complex surface phenomena*: For several reactants or for a reactant and a product or for transient adsorption, a superposition integral can be used for each solute species. These are coupled through the boundary conditions.

2. *Graetz problem*: This is a convective-diffusion example treated numerically in Chapter 12. After the solution for a jump change in surface concentration is worked out, even though it involves a series solution by separation of variables, the solution for the step change can be superposed in order to handle more complex boundary conditions, such as a prescribed concentration profile with axial distance or a heterogeneous catalyzed reaction.

3. *Inverse integral*: It is an integral where the surface concentration is expressed as an integral over the flux density at a surface. See Eq. 22.21.

4. *Intraparticle diffusion in the porous electrodes of a lithium-ion battery*: This gives improved computation speed if the solid-state diffusion coefficient can be regarded as independent of the composition.

5. *Mass transfer in thin diffusion layers, where the Lighthill transformation applies*: This is another example of a similarity transformation, and it is quite useful for a variety of geometries when the Schmidt number is large.

6. *Current sources along an electrode*: An example on this was touched upon in the preceding section. In these examples, with Laplace's equation, the upper limit may be L instead of t or x, reflecting the elliptic nature of the problem.

7. *A packed bed with or without axial diffusion*: Even in an experimental situation, it may be valuable to perceive the problem as one of generation from sources distributed along the surface of the packing.

8. *Diffusion from a cube (or a strip or a square)*: In statistical mechanics, a simulation may be made of the diffusion of

particles, or even of the steady fluctuations. To check on the accuracy of such calculations or to infer a value for the diffusion coefficient from the simulation, one may want for comparison the result of a diffusion problem. This was carried out in the chapter on Fourier transforms (Chapter 24). The section Additional Notes in that chapter, and some of the problems, point out that the solution can be regarded (perhaps more simply) as a superposition of diffusion from a distribution of point, plane, or line sources, whose form has been determined in a straightforward manner by a similarity transformation.

Problems

22.1 (a) If the inverse Laplace transform of $e^{-k\sqrt{s}}$ is

$$\frac{k}{2\sqrt{\pi t^3}}\exp\left(-\frac{k^2}{4t}\right),\qquad(22.32)$$

use the convolution integral to write down the inverse of Eq. 22.10.

(b) Derive the same result from Eq. 22.14 by integrating by parts.

22.2 Develop a reverse integral equation to that in Eq. 22.17 or 22.19 by applying again Laplace transforms. Differentiate Eq. 22.10 with respect to x and set $x = 0$ to obtain

$$G(s)=-\left(\frac{\alpha}{s}\right)^{1/2}H(s).\qquad(22.33)$$

Use the convolution integral to invert this with the help of the Laplace transform in Eq. 10.40. The result is in Eq. 22.21.

Additional Notes

Numerical Method for Volterra Integral Equations

Let us approach the problem in Eq. 22.27 by breaking τ up into a series of intervals of width $\Delta\tau$.

1. We already know the value at $\tau = 0$, namely, $\Theta = 1$.

2. Obtain the value at $\tau = \Delta\tau$. The values at $\tau = 2\delta\tau$, $3\delta\tau$, etc., cannot affect the result; the future does not directly affect the present. Let us worry about the details of this value of τ later.

3. Now skip to $\tau = k\Delta\tau$. Assume that Θ has been determined for $\tau \leq (k-1)\Delta\tau$. Thus, there is only one unknown to determine at a time.

4. In the numerical approach, take $d\Theta/d\tau$ to be constant over any time segment:

$$\frac{d\Theta}{d\tau} = \frac{\Theta_j - \Theta_{j-1}}{\Delta\tau}. \tag{22.34}$$

The integral equation becomes

$$-\Theta_k^2 = \frac{1}{\sqrt{\pi}} \sum_{j=2}^{k} \frac{\Theta_j - \Theta_{j-1}}{\Delta\tau} \int_{(j-2)\Delta\tau}^{(j-1)\Delta\tau} \frac{d\tau'}{\sqrt{(k-1)\Delta\tau - \tau'}}$$

$$= \frac{1}{\sqrt{\pi}} \sum_{j=2}^{k} \frac{\Theta_j - \Theta_{j-1}}{\Delta\tau} \frac{\sqrt{(k-1)\Delta\tau - \tau'}}{-1/2} \Bigg|_{(j-2)\Delta\tau}^{(j-1)\Delta\tau}$$

$$= \frac{1}{\sqrt{\pi}} \sum_{j=2}^{k} \frac{\Theta_j - \Theta_{j-1}}{\Delta\tau} (-2) \left[\sqrt{(k-1)-(j-1)} - \sqrt{(k-1)-(j-2)} \right] \Delta\tau^{1/2}$$

$$= \frac{-2}{\sqrt{\pi\Delta\tau}} \sum_{j=2}^{k} \left[\Theta_j - \Theta_{j-1} \right] \left[\sqrt{k-j} - \sqrt{k-j+1} \right] \tag{22.35}$$

Let

$$B_L = \sqrt{L} - \sqrt{L-1} = \frac{1}{\sqrt{L} + \sqrt{L-1}} \quad \text{for } L = 1, 2, \dots . \tag{22.36}$$

The equation needs to have the subscripts manipulated.

$$-\Theta_k^2 = \frac{2}{\sqrt{\pi\Delta\tau}} \sum_{j=2}^{k} \Theta_j B_{k-j+1} - \frac{2}{\sqrt{\pi\Delta\tau}} \sum_{j=2}^{k} \Theta_{j-1} B_{k-j+1} \tag{22.37}$$

or

$$-\frac{\sqrt{\pi\Delta\tau}}{2} \Theta_k^2 = \Theta_k B_1 + \sum_{j=2}^{k-1} \Theta_j B_{k-j+1} \sum_{J=2}^{k-1} \Theta_J B_{k-J}, \tag{22.38}$$

where in one sum we have let $J = j - 1$.

Note that $B_1 = 1$. Rewrite the second sum with J replaced by j, since it is a dummy variable.

$$\Theta_k + \frac{\sqrt{\pi \Delta \tau}}{2} \Theta_k^2 = \Theta_1 B_{k-1} + \sum_{j=2}^{k-1} \Theta_j \left[B_{k-j} - B_{k-j+1} \right]. \quad (22.39)$$

(The sum on the right is not included if $k = 2$.) The right side of the equation is known. Solve for Θ_k. Let

$$A_L = B_L - B_{L+1} = 2\sqrt{L} - \sqrt{L+1} - \sqrt{L-1}. \quad (22.40)$$

Then

$$\Theta_k + \frac{\sqrt{\pi \Delta \tau}}{2} \Theta_k^2 = \Theta_1 B_{k-1} + \sum_{j=2}^{k-1} \Theta_j A_{k-j}. \quad (22.41)$$

The integral equation is more efficient and accurate than a finite-difference solution. We have seen that we can handle a nonlinear boundary condition. The integral equation embodies the partial differential equation, the initial condition, and the boundary condition at $x = \infty$. The essential feature is that the differential equation is linear.

One can gain a little more accuracy for the first step. For short times, with

$$-\Theta^2 = \frac{1}{\sqrt{\pi}} \int_0^\tau \left. \frac{d\Theta}{d\tau} \right|_{\tau = \tau'} \frac{d\tau'}{\sqrt{\tau - \tau'}}, \quad (22.42)$$

and since $\Theta = 1$ at $\tau = 0$, we can try a solution that

$$\frac{d\Theta}{d\tau} = \frac{-A}{\sqrt{\tau}} \quad (22.43)$$

or

$$\Theta = 1 - 2A\sqrt{\tau} \quad \text{for small } \tau. \quad (22.44)$$

We can substitute into the integral equation to get a value for A:

$$-1 \approx \frac{-A}{\sqrt{\pi}} \int_0^\tau \frac{d\tau'}{\sqrt{\tau'}\sqrt{\tau - \tau'}}. \quad (22.45)$$

Let $y = \tau'/\tau$. This gives

$$1 = \frac{A}{\sqrt{\pi}} \int_0^1 \frac{dy}{\sqrt{y}\sqrt{1-y}} = A\sqrt{\pi}. \quad (22.46)$$

This gives $A = 1/\sqrt{\pi}$ and gives the proper approximation for Θ at the second mesh point:

$$\Theta \approx -2\sqrt{\tau/\pi}. \tag{22.47}$$

The computer program shows how to implement this algorithm for this problem in Fortran. The most essential parts of the algorithm are enclosed in a box. The graph shows the quantitative solution for Θ and Θ^2 as a function of τ, showing in detail what we had worked out qualitatively before even solving the problem. The interpretation remains the same; there is a period for short times where the rate is kinetically limited and $\Theta \approx 1$, and there is a long time period, which is mass-transfer limited and where Θ is small and the rate is inversely proportional to the square root of τ.

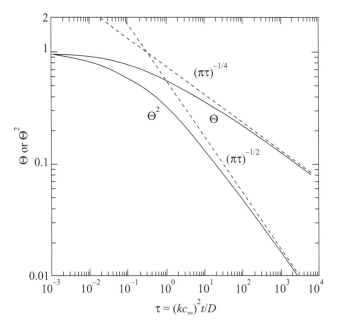

Figure 22.5 Surface concentration and heterogeneous reaction rate for diffusion from a semi-infinite stagnant medium with second-order reaction kinetics.

```
c      Program to solve for a surface reaction at the boundary
c      of a semi-infinite region, with a second-order
c      irreversible reaction. November 27, 2000.
       implicit real*8 (a-h,o-z)
       common th(201),is, Lmax,AA(201),BB(201),c2
       Lmax=201
       ex=0.5
       do 29 L=1,Lmax
       a=L
  29   BB(L) = 1.0/(a**ex+(a-1.0)**ex) ! = a**ex - (a-1.0)**ex
       do 30 L=1,Lmax-1
  30   AA(L)=BB(L)-BB(L+1)
       dt=0.001
       th(1)=1.0
       is=2
       do 40 kount=1,16
       c2=(3.141592654*dt)**ex
       call theta
       print 101, (dt*dble(j-1), th(j),j=is,Lmax)
 101   format (2x,f10.3, f12.5)
       do 35 j=1,Lmax,2
       is=(j-1)/2+1
  35   th(is)=th(j)
       dt=2.0*dt
  40   is=is + 1
       end
       subroutine theta
       implicit real*8 (a-h, o-z)
c      Subprogram for calculating concentration
       common th(201), is,Lmax,A(201),B(201),c2
```

```
       do 60 k=is,Lmax
       sum=th(1)*B(k-1)
       if(k.gt.2) then
       do 40 j=3,k
  40   sum=sum+th(j-1)*A(k-j+1)
       endif
  60   th(k)=2.0*sum/(1.0+(1.0 + 2.0*sum*c2)**0.5)
```

```
       return
       end
```

Figure 22.6 Computer program for second-order kinetics.

References

1. Carl Wagner, "On the numerical solution of Volterra integral equations," *Journal of Mathematics and Physics*, **32**, 289–301 (1954).
2. Andreas Acrivos and Paul L. Chambré, "Laminar boundary layer flows with surface reactions," *Industrial and Engineering Chemistry*, **49**, 1025–1029 (1957).

Chapter 23

Migration in Rapid Double-Layer Charging

If a component is charged, its flux must include a contribution from migration in an electric field:

$$\underline{N}_i = -D_i \nabla c_i + \left(z_i D_i F/RT\right)c_i\,\underline{E} + \underline{v}c_i\,. \qquad (23.1)$$

Here the "mobility" is $u_i = D_i/RT$ and represents the velocity attained (relative to the solution) under the action of a unit force per mole. The charge per mole is $z_i F$, and the electric force per mole is $z_i F\underline{E}$. Consequently, this gives rise to a migration velocity of $u_i z_i F\underline{E}$. Multiplication by the concentration gives the contribution to the flux. The conservation law still reads

$$\frac{\partial c_i}{\partial t} + \nabla\bullet\underline{N}_i = 0\,, \qquad (23.2)$$

in the absence of homogeneous reactions in the bulk of the fluid. Furthermore, the electric field is related to the charge density by Poisson's equation

$$\nabla\bullet\underline{E} = \rho_e\,/\,\varepsilon \qquad (23.3)$$

where

$$\rho_e = \sum z_i F c_i \qquad (23.4)$$

and ε is the permittivity of the fluid medium.

We want to consider the rapid charging of a double layer at a plane interface in a stagnant medium. The ions present are one kind of cation and one kind of anion, with equal and opposite charge

The Newman Lectures on Mathematics
John Newman and Vincent Battaglia
Copyright © 2018 Pan Stanford Publishing Pte. Ltd.
ISBN 978-981-4774-25-3 (Hardcover), 978-1-315-10885-8 (eBook)
www.panstanford.com

numbers, $z_+ = -z_- = z$. In rapid charging, diffusion can be neglected as a first approximation. The problem is then hyperbolic. We shall charge for a certain period of time at a constant current, so that the electric field approaches a constant, $-B$, at infinity. The anions are driven away from the electrode by the field and, in the absence of diffusion, leave a region where anions are virtually absent. If there is initially no excess of either anions or cations at the interface, this region will increase linearly in thickness with time with a velocity $z\_u\_FE = zD\_FB/RT$. In the region where anions are absent, cations will be driven, largely by their own charge, toward the electrode, where they will pile up, in the absence of diffusion.

Within this region of extended charge, the governing equations thus become

$$c_- = 0 \tag{23.5}$$

$$\frac{\partial c_+}{\partial t} = -\frac{zFD_+}{RT}\frac{\partial c_+E}{\partial x} \tag{23.6}$$

$$\frac{\partial E}{\partial x} = \frac{zF}{\varepsilon}c_+ . \tag{23.7}$$

The boundary conditions at $x = zDFBt/RT$ are

$$E = -B \tag{23.8}$$

$$c_+ = c_\infty. \tag{23.9}$$

Outside this region, the solution is simply

$$E = -B, c_+ = c_- = c_\infty. \tag{23.10}$$

We shall consider only the case where the diffusion coefficients are equal, $D_+ = D_- = D$, and we use the following dimensionless variables:

$$\tau = \frac{2c_\infty Dz^2F^2}{RT\varepsilon}t, \quad X = \frac{2c_\infty zF}{\varepsilon B}x, \quad \xi = \frac{E}{B}, \quad Q = \frac{c_+}{2c_\infty}. \tag{23.11}$$

This serves to eliminate the parameters, and the problem becomes

$$\frac{\partial Q}{\partial \tau} = -\frac{\partial Q\xi}{\partial X}, \quad \frac{\partial \xi}{\partial X} = Q, \tag{23.12}$$

with the boundary conditions

$$\xi = -1, \quad Q = \frac{1}{2} \text{ at } X = \tau . \tag{23.13}$$

Let us define characteristic coordinates α and β by

$$\left(\frac{\partial \tau}{\partial \alpha}\right)_\beta = 0 \quad \text{and} \quad \left(\frac{\partial X}{\partial \beta}\right)_\alpha = \xi\left(\frac{\partial \tau}{\partial \beta}\right)_\alpha. \qquad (23.14)$$

Thus, lines of constant β are also lines of constant τ, while lines of constant α have a slope $dX/d\tau$ in the X, τ plane equal to ξ and coincide with the trajectories of cations, since ξ is now the dimensionless velocity of the cations moving under the influence of the electric field.

The coordinate transformation takes the form

$$\frac{\partial \xi}{\partial \alpha} = \frac{\partial \xi}{\partial X}\frac{\partial X}{\partial \alpha} + \frac{\partial \xi}{\partial \tau}\frac{\partial \tau}{\partial \alpha} = Q\frac{\partial X}{\partial \alpha} \qquad (23.15)$$

$$\frac{\partial Q}{\partial \beta} = \frac{\partial Q}{\partial \tau}\frac{\partial \tau}{\partial \beta} + \frac{\partial Q}{\partial X}\frac{\partial X}{\partial \alpha} = \left(\frac{\partial Q}{\partial \tau} + \xi\frac{\partial Q}{\partial X}\right)\frac{\partial \tau}{\partial \beta} = -Q^2\frac{\partial \tau}{\partial \beta}. \qquad (23.16)$$

These, then, are the governing differential equations, and together with Eq. 23.14, they constitute four differential equations for determining the four quantities τ, X, ξ, and Q as functions of α and β.

We see that τ is an arbitrary function of β, for which we take

$$\beta = \tau. \qquad (23.17)$$

To remove the arbitrariness in α, we adopt the boundary condition $\alpha = \tau$ when $X = \tau$.

We should note that the characteristic curves were definite; it was only their labeling in terms of α and β that was arbitrary.

The characteristic equations now take the form

$$\frac{\partial Q}{\partial \beta} = \left(\frac{\partial Q}{\partial \tau}\right)_\alpha = -Q^2, \quad \frac{\partial \xi}{\partial \alpha} = Q\frac{\partial X}{\partial \alpha}, \quad \frac{\partial X}{\partial \beta} = \left(\frac{\partial X}{\partial \tau}\right)_\alpha = \xi. \qquad (23.18)$$

Integration of the first of these equations gives

$$Q = \frac{1}{\tau + 2 - \alpha}, \qquad (23.19)$$

satisfying the condition $Q = \frac{1}{2}$ at $\alpha = \tau$ (or at $X = \tau$). Then differentiation of the last of Eq. 23.18 with respect to α gives

$$\frac{\partial^2 X}{\partial \tau \partial \alpha} = \frac{\partial \xi}{\partial \alpha} = Q\frac{\partial X}{\partial \alpha} = \frac{1}{2 + \tau - \alpha}\frac{\partial X}{\partial \alpha}. \qquad (23.20)$$

Integration at constant α gives

$$\ln\frac{\partial X}{\partial\alpha} = \ln(2+\tau-\alpha)+\ln f(\alpha) \qquad (23.21)$$

or

$$\partial X / \partial\alpha = (2+\tau-\alpha)f(\alpha). \qquad (23.22)$$

Integration at constant τ gives

$$X = \int_{\tau}^{\alpha}(2+\tau-\alpha)f(\alpha)d\alpha +\tau, \qquad (23.23)$$

where the integration "constant" or function of τ was evaluated from the condition that $X = \tau$ at $\alpha = \tau$. Then

$$\xi = \frac{\partial X}{\partial\tau} = 1 + \int_{\tau}^{\alpha}f(\alpha)d\alpha - 2f(\tau). \qquad (23.24)$$

The condition $\xi = -1$ at $\tau = \alpha$ gives $f(\alpha) = 1$. Thus

$$\xi = \alpha - \tau - 1 \qquad (23.25)$$

$$X = -\frac{1}{2}\alpha^2 + (2+\tau)\alpha - \tau - \frac{1}{2}\tau^2. \qquad (23.26)$$

Solving for α in terms of X and τ, we obtain

$$\alpha = 2+\tau - \sqrt{4+2\tau - 2X} \qquad (23.27)$$

where the minus sign is selected for the square root so that $\alpha = \tau$ when $X = \tau$. In summary, the charge distribution is (see Fig. 23.1)

$$Q = \frac{1}{2+\tau-\alpha} = \frac{1}{\sqrt{4+2\tau-2X}}$$

and the field distribution is

$$\xi = 1 - \sqrt{4+2\tau-2X}.$$

The characteristic curves are sketched in Fig. 23.2, where the charging current was cut off at $\tau = 20$.

With reference to Fig. 23.1, during the charging period, Q is maintained at a value of 0.5 at $X = \tau$ by the arrival of cations from the bulk solution. The presence of the free charge density increases the electric field and hence increases the speed of the cations as they move through this region. This results in a decrease in the concentration of cations, since they arrive at the electrode surface sooner.

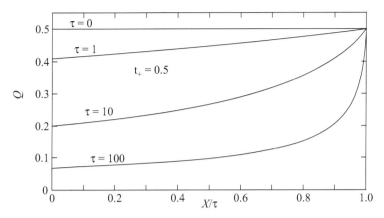

Figure 23.1 Distribution of extended charge during the charging period.

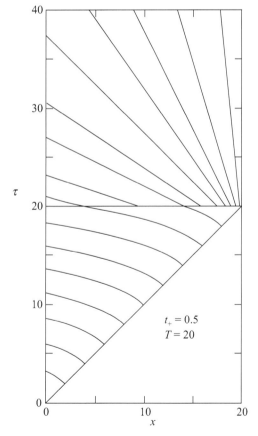

Figure 23.2 Characteristic curves for the charging and decay periods.

If the current is cut off at $\tau = T$, the subsequent decay of the extended charge can be treated in a similar manner with the results [1]

$$\xi = 2 + \tau' - \left[(2+\tau')^2 - (X-T)\right]^{1/2}$$

$$Q = 1 \Big/ \left[(2+\tau')^2 - (X-T)\right]^{1/2}$$

where $\tau' = \tau - T$. The charge distribution for the decay period is shown in Fig. 23.3. The region of extended charge now has a constant thickness corresponding to $X = T$, and there is no longer a supply of cations from the bulk solution.

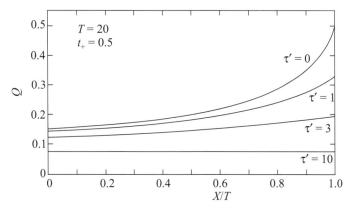

Figure 23.3 Distribution of extended charge after the charging period ($T=20$).

Reference

1. John Newman. "Migration in rapid double-layer charging." *Journal of Physical Chemistry*, **72**, 1843–1848 (1969).

Chapter 24

Fourier Transforms

Fourier transforms find scientific applications in fields such as spectroscopy. Some physical problems may be solved more readily with the Laplace transformation or by superposition integrals.

24.1 Finite Fourier Transform

Related to Fourier transforms is a subject called "finite Fourier transforms," sometimes abbreviated as FFT, although this should be reserved for "fast Fourier transform." The finite Fourier transform is really just the method of separation of variables, but with perhaps a different frame of mind gained by thinking about the Laplace transform or the Fourier transform.

Let us consider the problem (see Chapter 19)

$$\frac{\partial T}{\partial t} = \alpha \frac{\partial^2 T}{\partial x^2}, \tag{24.1}$$

with the boundary and initial conditions

$$T = 0 \quad \text{at} \quad x = 0, \tag{24.2}$$

$$\frac{\partial T}{\partial x} = 0 \quad \text{at} \quad x = L, \tag{24.3}$$

$$T = f(x) \quad \text{at} \quad t = 0. \tag{24.4}$$

The Newman Lectures on Mathematics
John Newman and Vincent Battaglia
Copyright © 2018 Pan Stanford Publishing Pte. Ltd.
ISBN 978-981-4774-25-3 (Hardcover), 978-1-315-10885-8 (eBook)
www.panstanford.com

The reasoning goes as follows: At any given time, we can expand the solution in a Fourier series using the functions

$$X_n = \sin \lambda_n x \quad \text{where} \quad \lambda_n = \left(n - \frac{1}{2}\right)\frac{\pi}{L}, n = 1, 2... \quad (24.5)$$

Thus writing

$$T = \sum_{n=1}^{\infty} P_n(t) X_n(x), \quad (24.6)$$

where we need to determine how the coefficients $P_n(t)$ vary with time. The fact that we can expand in this manner is already assured by the completeness of the set of functions.

Now multiply the governing equation by $X_n(x)$ and integrate (this is called, "taking the finite Fourier transform")

$$\int_0^L X_n \frac{\partial T}{\partial t} dx = \alpha \int_0^L X_n \frac{\partial^2 T}{\partial x^2} dx. \quad (24.7)$$

The operator $\partial^2 T/\partial x^2$ is supposed to be self adjoint so that one has

$$\frac{\partial}{\partial t} \int_0^L X_n T dx = \alpha \int_0^L T \frac{\partial^2 X_n}{\partial x^2} dx = -\alpha \int_0^L T \lambda_n^2 X_n dx. \quad (24.8)$$

Introduce Eq. 24.6 and use orthogonality to obtain

$$\frac{dP_n}{dt} \int_0^L X_n^2 dx = -\alpha \lambda_n^2 P_n \int_0^L X_n^2 dx \quad (24.9)$$

or

$$\frac{dP_n}{dt} = -\alpha \lambda_n^2 P_n. \quad (24.10)$$

Integration gives

$$P_n = C_n e^{-\alpha \lambda_n^2 t}. \quad (24.11)$$

Now treat the initial condition. Transform that also, that is, multiply by $X_n(x)$ and integrate.

$$\int_0^L T(x, 0) X_n dx = \int_0^L f(x) X_n dx. \quad (24.12)$$

Again from the orthogonality of the functions $X_n(x)$, we have

$$P_n(0)\int_0^L X_n^2 dx = P_n(0)\frac{L}{2}, \qquad (24.13)$$

giving

$$P_n(0) = \frac{2}{L}\int_0^L f(x)X_n dx. \qquad (24.14)$$

This gives the initial conditions for getting C_n (i.e., $C_n = P_n(0)$).

24.2 Extension to the Fourier Transform

By letting $L \to \infty$, we can go from the finite Fourier transform to the Fourier transform. Suppose that we have a function $f(x)$ defined from $x = 0$ to $x = L$, but nonzero only between 0 and a, where it has the value $f(x) = T_0$ for $0 \le x \le a$ (see Fig. 24.1). A function, over the whole domain, can be expressed as

$$f(x) = \frac{2}{\pi}\int_0^\infty B(\omega)\sin(\omega x)d\omega \qquad (24.15)$$

with

$$B(\omega) = \int_0^\infty f(x)\sin(\omega x)dx. \qquad (24.16)$$

The quantities f and B are referred to as a Fourier-sine-transform pair. You should note how the function $B(\omega)$ takes the place of the coefficients P_n in the example of the finite Fourier transform.

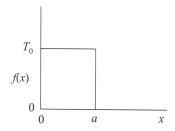

Figure 24.1 The initial temperature distribution is confined between 0 and a.

In the notation of that previous example, let us write

$$T(x,t) = \frac{2}{\pi} \int_0^\infty P(\omega,t)\sin(\omega x)d\omega. \qquad (24.17)$$

Multiply the differential equation 24.1 by $\sin(\omega x)dx$ and integrate ("taking the Fourier sine transform")

$$\frac{\partial}{\partial t}\int_0^\infty T\sin(\omega x)dx = \alpha \int_0^\infty \sin(\omega x)\frac{\partial^2 T}{\partial x^2}dx$$

$$= \alpha \int_0^\infty T\frac{\partial^2 \sin(\omega x)}{\partial x^2}dx = -\alpha\omega^2 \int_0^\infty T\sin(\omega x)dx. \quad (24.18)$$

From the definition of the Fourier sine transform,

$$P(\omega,t) = \int_0^\infty T(x,t)\sin(\omega x)dx, \qquad (24.19)$$

the above equation reduces to

$$\frac{\partial P(\omega,t)}{\partial t} = -\alpha\omega^2 P(\omega,t), \qquad (24.20)$$

with the solution

$$P(\omega,t) = P(\omega,0)e^{-\alpha\omega^2 t} \qquad (24.21)$$

$P(\omega, 0)$ should now be evaluated from the initial condition

$$P(\omega,0) = \int_0^\infty f(x)\sin(\omega x)dx. \qquad (24.22)$$

This example shows that it can be relatively straightforward to obtain the Fourier transform $P(\omega, t)$ for suitable problems. In analogy with the Laplace transform, it may be more difficult to obtain the desired function $T(x, t)$. It can be evaluated from the definition in Eq. 24.17. Note that in this example the Fourier sine transform was made with respect to the variable x, while with the Laplace transform one usually transforms with respect to the variable t.

24.3   More General Fourier Transforms

We developed a Fourier sine transform in the previous section, in the context of developing the solution in Chapter 19 when $L \to \infty$. The complete Fourier-transform pair

$$f(x) = \frac{1}{2\pi} \int_{-\infty}^{\infty} g(\omega) e^{i\omega x} d\omega \quad \text{and} \quad g(\omega) = \int_{-\infty}^{\infty} f(x) e^{-i\omega x} dx$$

$$(24.23)$$

applies over the infinite region $-\infty < x < \infty$ and is appropriate for functions that vanish at $x = \pm\infty$. (This can be considered an extension of Fourier series, which might arise by a solution by separation of variables.)

On a semiinfinite domain $0 < x < \infty$, we can have instead two alternative formulations: sines (which were developed earlier)

$$f(x) = \frac{2}{\pi} \int_0^{\infty} B(\omega) \sin(\omega x) d\omega \quad \text{with} \quad B(\omega) = \int_0^{\infty} f(x) \sin(\omega x) dx,$$

$$(24.24)$$

and cosine transform pairs

$$f(x) = \frac{2}{\pi} \int_0^{\infty} A(\omega) \cos(\omega x) d\omega \quad \text{with} \quad A(\omega) = \int_0^{\infty} f(x) \cos(\omega x) dx.$$

$$(24.25)$$

These can be regarded as odd and even parts of the complete Fourier transform, much as we developed Legendre Fourier series in a complete form for $-1 < x < 1$ and even and odd forms for $0 < x < 1$ (see Chapter 9). Also, note that the transform pairs are symmetric if x and ω are regarded as similar. The only difference is in the placement of the factor, $2/\pi$ or $1/2\pi$, and this placement is not critical as long as one is consistent.

24.4 Diffusion from a Finite Region

In statistical mechanics, fluctuations decay in the same way that material diffuses. (To see the similarity, the fluctuations should be correlated with the fluctuations at zero time, giving an *auto-correlation function*.) A dynamic Monte Carlo calculation or molecular dynamics simulates the fluctuations. The results can be compared to diffusion calculations, which we develop here.

We mention first one-dimensional diffusion from a finite region. The problem is to solve the differential equation

$$\frac{\partial T}{\partial t} = \alpha \frac{\partial^2 T}{\partial x^2}, \tag{24.26}$$

with the boundary and initial conditions

$$\frac{\partial T}{\partial x} = 0 \text{ at } x = 0, \tag{24.27}$$

$$T \to 0 \text{ as } x \to \infty, \tag{24.28}$$

$$\left.\begin{array}{ll} T = T_0 & \text{for } |x| \le a \\ T = 0 & \text{for } |x| > a \end{array}\right\} \text{ at } t = 0, \tag{24.29}$$

(see Fig. 24.2).

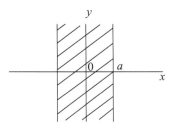

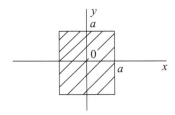

Figure 24.2 For one- and two-dimensional problems, the initial values are restricted as shown.

The two-dimensional analogue can be phrased as follows. Solve the differential equation

$$\frac{\partial T}{\partial t} = \alpha \left(\frac{\partial^2 T}{\partial x^2} + \frac{\partial^2 T}{\partial y^2} \right),$$ (24.30)

with the boundary and initial conditions

$$\frac{\partial T}{\partial x} = 0 \quad \text{at} \quad x = 0,$$ (24.31)

$$T \rightarrow 0 \quad \text{as} \quad x \rightarrow \infty,$$ (24.32)

$$\frac{\partial T}{\partial y} = 0 \quad \text{at} \quad y = 0,$$ (24.33)

$$T \rightarrow 0 \quad \text{as} \quad y \rightarrow \infty,$$ (24.34)

$$\left. \begin{array}{l} T = T_0 \text{ for } |x| < a \quad \text{and} \quad |y| < a \\ T = 0 \text{ otherwise} \end{array} \right\} \text{ at } t = 0.$$ (24.35)

This could also be done in three dimensions. We approach the problem in one dimension. Before going too much farther, you should consider that the problem might be easier to solve by Laplace transforms or by superposition integrals. These are covered in other chapters of this book.

Take the Fourier transform with respect to x.

$$\overline{T}(\omega, t) = \int_{-\infty}^{\infty} e^{-i\omega x} T(x, t) \, dx.$$ (24.36)

Taking the Fourier transform of both sides of the differential equation gives

$$\frac{\partial \overline{T}}{\partial t} = -\alpha \omega^2 \overline{T},$$ (24.37)

with the solution

$$\overline{T} = F(\omega) e^{-\alpha \omega^2 t},$$ (24.38)

analogous to Eq. 24.21. Remember that when integrating with respect to t at constant ω, the integration "constant" will be a function of ω.

To get $F(\omega)$, also take the Fourier transform of the initial condition:

$$\overline{T}(\omega,0) = F(\omega) = \int_{-a}^{a} e^{-i\omega x} T_0 \, dx = T_0 \left. \frac{e^{-i\omega x}}{-i\omega} \right|_{-a}^{a} = \frac{2T_0}{\omega} \sin(\omega a),$$

$$(24.39)$$

so that

$$\overline{T}(\omega,t) = \frac{2T_0}{\omega} \sin(\omega a) e^{-\alpha \omega^2 t}. \qquad (24.40)$$

The inverse Fourier transform could be calculated by direct application of the equation

$$T(x,t) = \frac{1}{2\pi} \int_{-\infty}^{\infty} e^{i\omega x} \overline{T}(\omega,t) \, d\omega. \qquad (24.41)$$

However, we choose instead to develop a more general tool, the convolution integral.

24.5   The Convolution Integral

Suppose that $\bar{c}_1(\omega)$ and $\bar{c}_2(\omega)$ are the Fourier transforms of two functions $c_1(x)$ and $c_2(x)$, so that, according to the definitions of the transform pair in Eq. 24.23

$$\bar{c}_1(\omega)\bar{c}_2(\omega) = \int_{-\infty}^{\infty} e^{-i\omega x} c_1(x) \, dx \int_{-\infty}^{\infty} e^{-i\omega x} c_2(x) \, dx. \qquad (24.42)$$

In the first integral, replace the dummy variable x by x', and combine the integrals.

$$\bar{c}_1(\omega)\bar{c}_2(\omega) = \int_{-\infty}^{\infty} e^{-i\omega x'} c_1(x') \, dx' \int_{-\infty}^{\infty} e^{-i\omega x} c_2(x) \, dx.$$

$$= \int_{-\infty}^{\infty} \int_{-\infty}^{\infty} e^{-i\omega (x'+x)} c_1(x') c_2(x) \, dx' \, dx. \qquad (24.43)$$

Replace x' by $X - x$.

$$\bar{c}_1(\omega)\bar{c}_2(\omega) = \int_{-\infty}^{\infty} \int_{-\infty}^{\infty} e^{-i\omega X} c_1(X - x) \, c_2(x) \, dX \, dx. \qquad (24.44)$$

Interchange of the order of integration gives

$$\bar{c}_1(\omega)\bar{c}_2(\omega) = \int\limits_{-\infty}^{\infty} e^{-i\omega X} \int\limits_{-\infty}^{\infty} c_1(X-x)c_2(x)\,dx\,dX. \qquad (24.45)$$

Let us use the notation $\mathcal{F}$ for the Fourier transform and $\mathcal{F}^{-1}$ for the inverse transform. Then, we have established that

$$\mathcal{F}^{-1}[\bar{c}_1(\omega)\bar{c}_2(\omega)] = \int\limits_{-\infty}^{\infty} c_1(x-X)c_2(X)\,dX. \qquad (24.46)$$

The integral on the right is called the convolution of $c_1(x)$ and $c_2(x)$ and permits the inverse of the product of two Fourier transforms to be written down.

24.6 Finishing the Diffusion Problem

For the diffusion problem, before applying the specific initial condition, we obtained Eq. 24.38. This can be regarded as the product of two Fourier transforms. Before we can apply the convolution theorem, we must establish a Fourier transform pair. Let us evaluate the integral

$$I(x) = \int\limits_{-\infty}^{\infty} e^{i\omega x} e^{-\alpha\omega^2 t}\,d\omega. \qquad (24.47)$$

We can complete the square to express this as

$$I(x) = \int_{-\infty}^{\infty} \exp\left[-\alpha t\left(\omega - \frac{ix}{2\alpha t}\right)^2 - \frac{x^2}{4\alpha t}\right] d\omega$$

$$= \exp\left(-\frac{x^2}{4\alpha t}\right)\int_{-\infty}^{\infty} \exp\left[-\alpha t\left(\omega - \frac{ix}{2\alpha t}\right)^2\right] d\omega. \qquad (24.48)$$

This can be rewritten as

$$I(x) = \exp\left(-\frac{x^2}{4\alpha t}\right)\int_{-\infty}^{\infty} \exp(-\alpha t\Omega^2)\,d\Omega, \qquad (24.49)$$

by letting $\Omega = \omega - ix/2\alpha t$. By writing $y = \Omega\sqrt{\alpha t}$, we obtain

$$I(x)=\left(\frac{1}{\sqrt{\alpha t}}\right)\exp\left(\frac{-x^2}{4\alpha t}\right)\int_{-\infty}^{\infty}e^{-y^2}dy=\left(\frac{\pi}{\alpha t}\right)^{1/2}\exp\left(\frac{-x^2}{4\alpha t}\right). \quad (24.50)$$

Another way to say this is that

$$\mathcal{F}\left[\frac{1}{2\sqrt{\pi\alpha t}}\exp\left(-\frac{x^2}{4\alpha t}\right)\right]=e^{-\alpha\omega^2 t}. \quad (24.51)$$

The gaussian form in terms of x yields the gaussian form in terms of ω, a remarkable result.

With this new Fourier transform pair, we can apply the convolution theorem to the Fourier transform in Eq. 24.38. We obtain

$$T(x,t)=\int_{-\infty}^{\infty}\frac{f(x-X)}{2\sqrt{\pi\alpha t}}\exp\left(\frac{-X^2}{4\alpha t}\right)dX$$

$$=\frac{1}{2\sqrt{\pi\alpha t}}\int_{-\infty}^{\infty}f(X)\exp\left(\frac{-(x-X)^2}{4\alpha t}\right)dX. \quad (24.52)$$

We write this in two ways in order to emphasize that they both are the same. But we also want to interpret this as a source at X which is then diffusing in time and space according to the gaussian solution measured from X, as in $x - X$, while decaying in time like $1/\sqrt{t}$.

Now it is relatively easy to insert the specific form for $f(x)$ applying in this problem, as depicted in Fig. 24.2 and expressed in Eq. 24.29.

$$T(x,t)=\frac{T_0}{2\sqrt{\pi\alpha t}}\int_{-a}^{a}\exp\left(\frac{-(x-X)^2}{4\alpha t}\right)dX. \quad (24.53)$$

If we want to know how much material remains in the original region $-a < x < a$, we need to integrate this again. This should eventually be able to be reduced to a single error function.

The similarity of Fourier transforms to Laplace transforms should be clear by now. It is expedient in many cases to determine transform pairs and to determine an inverse (Fourier or Laplace) transform by looking up the desired result in a table. It is also useful to establish certain general theorems or formulas, of which the convolution integral is a good example. A formula for differentiating is useful because it helps us solve differential equations with a simple methodology. We saw this when we transformed the diffusion equation. (See also Problems 24.3 and 24.4)

24.7 Formal Connection with Laplace Transforms

Let us express the transform pair of Eq. 24.23 with x replaced by t:

$$f(t) = \frac{1}{2\pi} \int_{-\infty}^{\infty} g(\omega) e^{i\omega t} d\omega \quad \text{and} \quad g(\omega) = \int_{-\infty}^{\infty} f(t) e^{-i\omega t} dt. \quad (24.54)$$

We could contemplate whether $g(\omega)$ should be real if $f(t)$ is real, because we associate a Fourier transform with a real quantity. In any case, let us explore the connection with Laplace transforms by first requiring $f(t)$ to be zero for $t < 0$.

$$g(\omega) = \int_{0}^{\infty} f(t) e^{-i\omega t} dt. \quad (24.55)$$

Next, let $i\omega = s$ and $g(\omega) = F(s)$. Then we obtain, first, the definition of the Laplace transform (see Eq. 10.1)

$$F(s) = \int_{0}^{\infty} f(t) e^{-st} dt. \quad (24.56)$$

The other half of the pair becomes, formally, the formula for the inverse Laplace transformation (see Eq. 20.1)

$$f(t) = \frac{1}{2\pi i} \lim_{\beta \to \infty} \int_{\gamma - i\beta}^{\gamma + i\beta} F(s) e^{st} ds, \quad (24.57)$$

where γ is large enough for any nonanalytic behavior of $F(s)$ to lie to the left of the path of integration. γ is zero in the original transform pair but was allowed to vary in the last equation to achieve a greater generality. (See Fig. 24.3.) This will be the closest thing to the derivation of the inverse Laplace transform formula (used in Chapter 20) that you will see in this course. The Laplace transformation is more powerful than the Fourier transform in a sense, because functions $f(t)$ do not need to be bounded at infinity in order for the Laplace transform to exist. Recall that an analytic function always yields zero when integrated around a closed contour. Therefore it is possible to move the value of γ within a certain range without changing the value of the integral in Eq. 20.1.

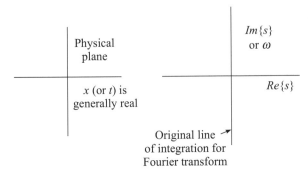

Figure 24.3   Integration paths for Laplace and Fourier transforms.

Problems

24.1 (a) Take Eq. 24.16 to be the definition of the Fourier sine transform. Prove the direct transform of Eq. 24.15 by direct calculations. Do this by assuming Eq. 24.15 is valid. Substitute Eq. 24.15 in, reverse the order of integration, and reduce this to an identity ($f(x) = f(x)$).

   (b) Prove the same result by paralleling the finite Fourier transform (i.e., determine whether there is a step similar to using orthogonality).

24.2 Contemplate the answer in Eq. 24.53.

   (a) What do we obtain for $T(x, t)$ when $t \to 0$?

   (b) What do we obtain for $T(x, t)$ when t is large? Describe the result in words.

24.3 Develop the transform of a derivative. Justify (or derive) with integration by parts and use of the property of $T(\to 0$ as $x \to \pm\infty)$. Note the analogy with the Laplace transform.

24.4 (a) Show what self adjoint means by integrating Eq. 24.7 by parts and using the properties of $X_n(x)$.

   (b) For Eq. 24.18, do we need to use the property that $T \to 0$ as $x \to \pm\infty$?

Additional Notes

Identifying Transform Pairs

Earlier we started with the formula 24.47 and derived the formula 24.50. Consistent with Eq. 24.23, we can express this as a transform pair

$$f(x) = \frac{e^{-x^2/4\alpha t}}{2\sqrt{\pi \alpha t}} \quad \text{and} \quad g(\omega) = e^{-\alpha \omega^2 t}. \tag{24.58}$$

Similarly, we derived the convolution formula 24.45

$$\bar{c}_1(\omega)\bar{c}_2(\omega) = \int_{-\infty}^{\infty} e^{-i\omega x} \int_{-\infty}^{\infty} c_1(X) c_2(x - X)\, dX dx. \tag{24.59}$$

Consistent with Eq. 24.23, we can express this as a transform pair

$$f(x) = \int_{-\infty}^{\infty} c_1(X) c_2(x - X)\, dX dx \quad \text{with } g(\omega) = c_1(\omega)\, c_2(\omega).$$

$$\tag{24.60}$$

It is important to include consistently the factor of 2π from Eq. 24.23.

Similarity Transformation and Superposition Integral

As a check on the result for diffusion from a box, we should show that a similarity transformation can give us a basic solution for diffusion from a plane and that this can be used with a superposition integral to get the same result as that given in Eq. 24.52.

For the differential equation

$$\frac{\partial T_1}{\partial t} = \alpha \frac{\partial^2 T_1}{\partial x^2}, \tag{24.61}$$

with the boundary and initial conditions

$$T_1 \to 0 \text{ as } x \to \pm\infty, \tag{24.62}$$

$$\left. \begin{array}{l} \int_{-\infty}^{\infty} T_1 dx = 1 \\[2mm] T_1 = 0 \quad \text{if} \quad x \neq 0 \end{array} \right\} \quad \text{at} \quad t = 0, \tag{24.63}$$

we try a solution of the form

$$T_1 = h(t)\Theta(\eta), \quad \text{where} \quad \eta = x/g(t). \quad (24.64)$$

Proceeding as we have been taught for similarity transformations of this type, we obtain the differential equation

$$h'\Theta - h\Theta'\eta\frac{g'}{g} = \alpha\frac{h\Theta''}{g^2}, \quad (24.65)$$

or

$$\Theta'' + \frac{gg'}{\alpha}\eta\Theta' - \frac{h'g^2}{\alpha h}\Theta = 0. \quad (24.66)$$

Here the primes denote differentiation with respect to either η or t, depending upon which argument is appropriate. For the similarity transformation to be successful, the coefficients in this equation must be independent of t. Set the first coefficient to 2, obtaining the differential equation for $g(t)$

$$g\frac{dg}{dt} = 2\alpha, \quad (24.67)$$

with the solution

$$g = 2\sqrt{\alpha t}, \quad (24.68)$$

after we have set the constant of integration to zero so that the initial conditions can be satisfied.

We set the second coefficient to K, since we do not yet know its value. This gives us the differential equation for $h(t)$:

$$\frac{h'g^2}{\alpha h} = K, \quad (24.69)$$

giving the solution

$$\ln h = \frac{K}{4}\ln t + \ln D \quad \text{or} \quad h = Dt^{K/4}, \quad (24.70)$$

where the integration constant is not yet specified and the above solution for $g(t)$ has been used.

To get the value of K, we use the integral in the initial condition. This becomes (after using the transformation)

$$\int_{-\infty}^{\infty} h(t)\Theta(\eta)g(t)d\eta = 1. \quad (24.71)$$

Therefore, the product gh must be a constant, and the constant K above takes on the value -2. Take $h(t) = 1/g(t) = 1/2\sqrt{\alpha t}$.

The differential equation governing $\Theta(\eta)$ now takes the form

$$\Theta'' + 2\eta\Theta' + 2\Theta = 0, \tag{24.72}$$

with boundary conditions that

$$\Theta \to 0 \text{ as } x \to \pm\infty \tag{24.73}$$

and the integral under the curve is specified

$$\int_{-\infty}^{\infty} \Theta d\eta = 1. \tag{24.74}$$

The differential equation for Θ is linear and second order. We find that $e^{-\eta^2}$ is one solution of the homogeneous equation. In this manner we work out that the solution, including the normalization factor, is

$$\Theta = \frac{e^{-\eta^2}}{\sqrt{\pi}} \tag{24.75}$$

or

$$T_1 = \frac{1}{2\sqrt{\pi\alpha t}} \exp\left(\frac{-x^2}{4\alpha t}\right). \tag{24.76}$$

This is the similarity solution for diffusion from a plane (at $x = 0$) into the surrounding medium. If, at $t = 0$, we have a distribution of such sources, distributed with strength $f(x)$, then a superposition integral can be used to express the simultaneous diffusion from all these sources. This result is identical to the convolution result in Eq. 24.52.

Chapter 25

Conformal Mapping

We have already studied coordinate transformations. Conformal mapping is a particular type of coordinate transformation—one which is related intimately to complex variables.

25.1   A Coordinate Transformation

Let us start with the problem of heat conduction within a rectangle. Then the temperature satisfied Laplace's equation in two rectangular directions (see Chapter 15)

$$\frac{\partial^2 T}{\partial x^2} + \frac{\partial^2 T}{\partial y^2} = 0. \tag{25.1}$$

We let the height of the rectangle, W, approach infinity, so that no boundary condition is necessary there (or we can see by inspection that $T \to 0$ as $y \to \infty$). The left boundary was an insulator

$$\frac{\partial T}{\partial x} = 0 \quad \text{at} \quad x = 0. \tag{25.2}$$

The right boundary was at zero temperature, while the bottom boundary was at a fixed temperature T_0

$$T = 0 \quad \text{at} \quad x = L \tag{25.3}$$

The Newman Lectures on Mathematics
John Newman and Vincent Battaglia
Copyright © 2018 Pan Stanford Publishing Pte. Ltd.
ISBN 978-981-4774-25-3 (Hardcover), 978-1-315-10885-8 (eBook)
www.panstanford.com

and

$$T = T_0 \quad \text{at} \quad y = 0. \tag{25.4}$$

We had particular problems because of the discontinuity of temperature at the lower right corner. (Figure 25.1 includes a sketch of this situation.)

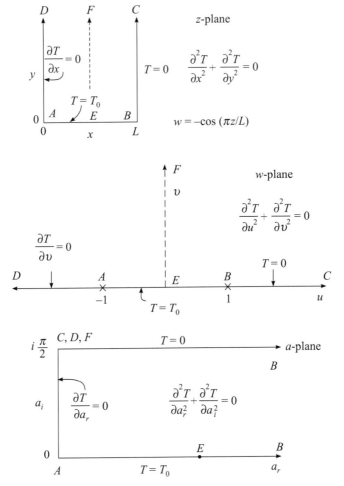

Figure 25.1 Conformal mapping of a heat-conduction problem through three configurations.

Now consider x and y to be the real and imaginary parts of a complex variable z

$$z = x + iy, \tag{25.5}$$

and consider the function

$$w = f(z) = \sin\left(\frac{\pi z}{L} - \frac{\pi}{2}\right) = -\cos\left(\pi\frac{z}{L}\right). \tag{25.6}$$

This is an ordinary function which we should expect to be able to differentiate as many times as we want. Except for the need to interpret everything as functions of complex variables, we would proceed just as if we were dealing with real functions.

The first thing we should like to do with the function w is to regard it as a mapping of points from the z-plane into points in the w-plane. In the z-plane, we are interested mainly in points with values of x between 0 and L. On the real axis, such points are mapped into points on the real w-axis between –1 and +1. If we continued on the x-axis beyond $x = L$, we would find ourselves oscillating on the real w-axis, always being between –1 and +1. This is because the sine function is periodic. It also means that the reverse mapping is multivalued, and one can find many points on the x-axis that correspond to a given point, say $w = 0$, on the real w-axis between –1 and +1. (Figure 25.1 includes the w-plane.)

Next follow the y-axis in the mapping. We get

$$w = -\cos\left(\frac{i\pi y}{L}\right) = -\cos h\left(\frac{\pi y}{L}\right). \tag{25.7}$$

Thus the y-axis maps into the negative real w-axis, to the left of –1. Similarly, the line at $z = L + iy$ maps into the positive real w-axis, to the right of +1. We can infer that our original region of interest maps into the upper half of the w-plane. Closer examination would show that angles are preserved in the mapping, except at those points where $f'(z) = 0$. Those are the points where $\sin(\pi z/L) = 0$, two of which are the lower corners of our region of interest, namely $x = 0$ and $x = L$ at $y = 0$. We already observed that at these points we had distorted the angle by mapping three lines in the z-plane all into the real w-axis.

We can continue to map out the transformation, taking other points in the region of interest and seeing where they lie in the *w*-plane. The result will bear a strong resemblance to the sketch shown in Fig. 16.1, although there we dealt with an axisymmetric system and here we deal with a cylindrical system, extending uniformly in the direction perpendicular to the *z*-plane. Figure 25.2 shows lines of constant *y* and constant *x*, plotted in the *w*-plane.

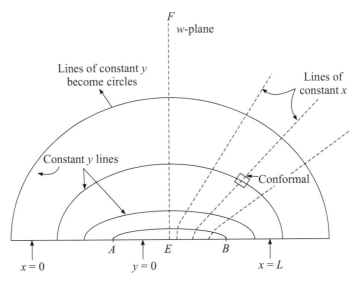

Figure 25.2 Lines of constant *x* and *y*, plotted in the *w*-plane. Compare with Fig. 16.1. Here we have elliptic coordinates, a two-dimensional system. Rotating about the *EF* line give rotational elliptic coordinates or oblate spheroidal coordinates, an axisymmetric system (see Chapters 16–18). Rotating about the *AB* line gives prolate spheroidal coordinates, another axisymmetric system.

You might well ask what advantage we have gained by mapping the original problem into the *w*-plane—after all, we still need to develop the coordinate transformation, find the governing partial differential equation in the *w*-plane, and solve it subject to boundary conditions yet to be stated in the *w*-plane. It does not look as though solving the problem in the *w*-plane is any easier than solving it in the *z*-plane, and indeed this observation is correct. We will need to do another transformation, from the *w*-plane, before the solution for the temperature would become obvious. Even after that, we might need to transform the result back into the original coordinate system. (Figure 25.1 also shows how things will look in the *a*-plane.)

25.2 Analytic Functions

If a complex function $f(z)$ has a continuous derivative $f'(z)$ at every point within a region, then the function is said to be analytic within the region and to be analytic at each point within the region. (Thus, at any point where the function is analytic, there must be surrounding points where the function is also analytic.) To be continuous, the derivative must also give the same value at a point, no matter how the derivative is calculated. Let

$$f(z) = u + iv. \qquad (25.8)$$

The derivative is defined as

$$\frac{df}{dz} = f'(z) = \lim_{\Delta z \to 0} \frac{\Delta f}{\Delta z} = \lim_{\Delta z \to 0} \frac{\Delta u + i \Delta v}{\Delta x + i \Delta y}. \qquad (25.9)$$

The requirement that the derivative be continuous means that we obtain the same results no matter how we let Δz approach zero. First let Δy be zero while we let $\Delta x \to 0$

$$\frac{df}{dz} = \lim_{\Delta z \to 0} \frac{\Delta u}{\Delta x} + i \frac{\Delta v}{\Delta x} = \frac{\partial u}{\partial x} + i \frac{\partial v}{\partial x}. \qquad (25.10)$$

If instead we let Δx be zero while we let $\Delta y \to 0$, we obtain

$$\frac{df}{dz} = \lim_{\Delta z \to 0} \frac{\Delta u}{i \Delta y} + i \frac{\Delta v}{i \Delta y} = \frac{\partial v}{\partial y} - i \frac{\partial u}{\partial y}. \qquad (25.11)$$

Equating the real and imaginary parts of these two expressions yields the Cauchy–Riemann conditions

$$\frac{\partial u}{\partial x} = \frac{\partial v}{\partial y} \text{ and } \frac{\partial v}{\partial x} = -\frac{\partial u}{\partial y}, \qquad (25.12)$$

prevailing for the real and imaginary parts of an analytic function. We have also obtained two expressions for the derivative:

$$f'(z) = \frac{\partial u}{\partial x} + i \frac{\partial v}{\partial x} = \frac{\partial v}{\partial y} - i \frac{\partial u}{\partial y}. \qquad (25.13)$$

The logic works in the reverse direction also; if the Cauchy–Riemann conditions are satisfied, then u and v constitute the real and imaginary parts of an analytic function of the complex variable $z = x + iy$.

We can use the Cauchy–Riemann conditions to show that the real and imaginary parts of an analytic function must each satisfy Laplace's equation as functions of x and y. Differentiate both of the Cauchy–Riemann conditions with respect to both x and y to obtain

$$\frac{\partial^2 u}{\partial x^2} = \frac{\partial^2 v}{\partial x \partial y}, \tag{25.14}$$

$$\frac{\partial^2 u}{\partial y \partial x} = \frac{\partial^2 v}{\partial y^2}, \tag{25.15}$$

$$\frac{\partial^2 v}{\partial x^2} = -\frac{\partial^2 u}{\partial x \partial y}, \tag{25.16}$$

$$\frac{\partial^2 v}{\partial y \partial x} = -\frac{\partial^2 u}{\partial y^2}. \tag{25.17}$$

Addition of Eqs. 25.14 and 25.17 gives

$$\frac{\partial^2 u}{\partial x^2} + \frac{\partial^2 u}{\partial y^2} = 0 \tag{25.18}$$

since the second cross derivatives cancel. Similarly, subtraction of Eqs. 25.15 and 25.16 gives

$$\frac{\partial^2 v}{\partial x^2} + \frac{\partial^2 v}{\partial y^2} = 0. \tag{25.19}$$

Functions, like u and v, which satisfy Laplace's equation are called harmonic. Since u and v are related by being real and imaginary parts of an analytic function, they are further called conjugate harmonic functions. For a given harmonic function, like u, one can find the conjugate harmonic function, except for an additive constant, by integrating the Cauchy–Riemann conditions.

We have now given you an infinite number of ways of constructing functions that satisfy Laplace's equation, namely by looking at the real and imaginary parts of any analytic function. Recall that in Chapter 14, we determined the general solution of Laplace's equation. We stated that we had never seen any real application of that general solution. As a challenge, see whether you can establish a relationship between the function $f(z)$ and its real and imaginary parts, u and v, and the general solution stated in Chapter 14 (see Problem 25.2).

25.3   Conformal Mapping

A conformal mapping is now defined as the one resulting from an analytic function, $f(z)$. The word conformal means here that a small part of the transformation retains its shape, that is, angles are preserved. Right angles remain right angles (except where $f'(z) = 0$). The transformation can be stretched and rotated by multiplying by a constant. A complex constant will rotate the figure, while a real constant merely expands or contracts it. The transformation can be translocated by adding a constant to it. This effectively moves the origin of the new coordinate system. (A little later we will prove that angles are preserved by a conformal mapping.)

As we shall see, a conformal mapping preserves equipotential surfaces, and insulators remain insulators. The insulators, or current lines, are perpendicular to the equipotential surfaces at all points; the insulators can be the surfaces of constant values of a harmonic function, and the equipotentials are surfaces of constant values of the conjugate harmonic function.

25.4   Preserving Laplace's Equation

We now view the conformal mapping as a coordinate transformation from x, y to u, v. What is the governing equation in terms of u and v?

To show that angles are preserved by the transformation, write

$$df = \frac{df}{dz}dz = \frac{df}{dz}(dx + idy) = du + idv. \qquad (25.20)$$

Therefore,

$$(\Delta u)^2 + (\Delta v)^2 = \left|\frac{df}{dz}\right|^2 \left[(\Delta x)^2 + (\Delta y)^2\right]. \qquad (25.21)$$

Here, $|df/dz|$ is a magnification factor. Also, the angle of $\Delta v/\Delta u$ = angle of $\Delta y/\Delta x$ + angle of df/dz. Thus, angles between two lines in the z-plane are preserved in the w-plane; a small image in the z-plane preserves its shape; it is rotated by the angle of df/dz; and it is magnified by the factor $|df/dz|$. Generally it is also translated.

Laplace's equation is also preserved in the mapping. This is a specific example of a coordinate transformation $(x, y \rightarrow u, v)$. What happens to Eq. 25.1 in this transformation? This is a general problem we need to face often.

From the definition $w = f(z)$, we can infer that $u = u(x, y)$ and $v = v(x, y)$ are known. The total differential of T is

$$dT = \frac{\partial T}{\partial u} du + \frac{\partial T}{\partial v} dv. \tag{25.22}$$

Hence,

$$\frac{\partial T}{\partial x} = \frac{\partial T}{\partial u}\frac{\partial u}{\partial x} + \frac{\partial T}{\partial v}\frac{\partial v}{\partial x} \tag{25.23}$$

and

$$\frac{\partial T}{\partial y} = \frac{\partial T}{\partial u}\frac{\partial u}{\partial y} + \frac{\partial T}{\partial v}\frac{\partial v}{\partial y}. \tag{25.24}$$

Chapter 16 recommends that you always carry a subscript on a derivative to tell you what is being held constant. Otherwise you are likely to be confused. We can get away with it here because we have been careful, and perhaps lucky.

Next we need $\partial^2 T/\partial x^2$ and $\partial^2 T/\partial y^2$.

$$\frac{\partial^2 T}{\partial x^2} = \frac{\partial u}{\partial x}\left[\frac{\partial^2 T}{\partial u^2}\frac{\partial u}{\partial x} + \frac{\partial^2 T}{\partial v \partial u}\frac{\partial v}{\partial x}\right] + \frac{\partial^2 u}{\partial x^2}\frac{\partial T}{\partial u}$$
$$+ \frac{\partial v}{\partial x}\left[\frac{\partial^2 T}{\partial u \partial v}\frac{\partial u}{\partial x} + \frac{\partial^2 T}{\partial v^2}\frac{\partial v}{\partial x}\right] + \frac{\partial^2 v}{\partial x^2}\frac{\partial T}{\partial v}. \tag{25.25}$$

Similarly,

$$\frac{\partial^2 T}{\partial y^2} = \frac{\partial u}{\partial y}\left[\frac{\partial^2 T}{\partial u^2}\frac{\partial u}{\partial y} + \frac{\partial^2 T}{\partial v \partial u}\frac{\partial v}{\partial y}\right] + \frac{\partial^2 u}{\partial y^2}\frac{\partial T}{\partial u}$$
$$+ \frac{\partial v}{\partial y}\left[\frac{\partial^2 T}{\partial u \partial v}\frac{\partial u}{\partial y} + \frac{\partial^2 T}{\partial v^2}\frac{\partial v}{\partial y}\right] + \frac{\partial^2 v}{\partial y^2}\frac{\partial T}{\partial v}. \tag{25.26}$$

Now add, grouping terms where possible.

$$\frac{\partial^2 T}{\partial x^2} + \frac{\partial^2 T}{\partial y^2} = \frac{\partial^2 T}{\partial u^2}\left[\left(\frac{\partial u}{\partial x}\right)^2 + \left(\frac{\partial u}{\partial y}\right)^2\right] + \frac{\partial^2 T}{\partial v \partial u}\left[2\frac{\partial u}{\partial x}\frac{\partial v}{\partial x} + 2\frac{\partial u}{\partial y}\frac{\partial v}{\partial y}\right]$$
$$+ \frac{\partial^2 T}{\partial v^2}\left[\left(\frac{\partial v}{\partial x}\right)^2 + \left(\frac{\partial v}{\partial y}\right)^2\right] + \frac{\partial T}{\partial u}\left[\frac{\partial^2 u}{\partial x^2} + \frac{\partial^2 u}{\partial y^2}\right] + \frac{\partial T}{\partial v}\left[\frac{\partial^2 v}{\partial x^2} + \frac{\partial^2 v}{\partial y^2}\right]. \tag{25.27}$$

Since u and υ are harmonic, the last two terms vanish. By the Cauchy–Riemann conditions, Eqs. 25.12, the coefficient of the cross derivative vanishes,

$$2\frac{\partial \upsilon}{\partial y}\frac{\partial \upsilon}{\partial x} - 2\frac{\partial \upsilon}{\partial x}\frac{\partial \upsilon}{\partial y} = 0.$$

The coefficients of the remaining two terms are equal.

$$\frac{\partial^2 T}{\partial x^2} + \frac{\partial^2 T}{\partial y^2} = \left|\frac{df}{dz}\right|^2 \left(\frac{\partial^2 T}{\partial u^2} + \frac{\partial^2 T}{\partial \upsilon^2}\right), \qquad (25.28)$$

where

$$\left|\frac{df}{dz}\right|^2 = \left(\frac{\partial u}{\partial x}\right)^2 + \left(\frac{\partial u}{\partial y}\right)^2 = \left(\frac{\partial \upsilon}{\partial x}\right)^2 + \left(\frac{\partial \upsilon}{\partial y}\right)^2. \qquad (25.29)$$

(This is the square of the magnification factor $|df/dz|$.) For Laplace's equation, this means that

$$\frac{\partial^2 T}{\partial x^2} + \frac{\partial^2 T}{\partial y^2} = 0 \text{ becomes}$$

$$\frac{\partial^2 T}{\partial u^2} + \frac{\partial^2 T}{\partial \upsilon^2} = 0 \qquad (25.30)$$

for a conformal mapping.

To summarize, some of the nice features of any conformal mapping are

1. Laplace's equation is preserved.
2. Lines of constant T in the z-plane remain lines of constant T in the w-plane. (This is belaboring the obvious.)
3. Insulators in the z-plane remain insulators in the w-plane. That is, if $\partial T/\partial n = 0$ in the z-plane, $\partial T/\partial n' = 0$ in the w-plane, where n' remains perpendicular to the surface. (Recall that angles are preserved.)
4. Angles are preserved, angle Δu, $\Delta \upsilon$ = angle Δx, Δy + angle df/dz, unless $df/dz = 0$ as at A and B.
5. The local amplification factor is $|df/dz|$. That is,

$$\sqrt{(\Delta u)^2 + (\Delta \upsilon)^2} = \left|\frac{df}{dz}\right|\sqrt{(\Delta x)^2 + (\Delta y)^2}. \qquad (25.31)$$

25.5   The Schwarz–Christoffel Transformation

Before we can complete the heat-conduction problem introduced at the beginning of the chapter (as carried over from Chapter 15), we need to introduce one more transformation, from the w-plane to what we call the a-plane. We want the problem in the a-plane to consist of solving Laplace's equation in a region where two opposite sides of a rectangular region are at temperatures T_0 and 0, respectively, and where one end of the region has an insulator boundary condition and the other end extends to infinity. (See the a-plane in Fig. 25.1.) But the transformation we want to use is an example of a fairly general transformation whereby the upper half plane (as in the w-plane) is mapped into the interior of a polygon. The reverse transformation from the w-plane to the z-plane is an example.

The Schwarz–Christoffel transformation has the general form

$$a = a(w) = C \int^{w} (w - u_1)^{k_1} (w - u_2)^{k_2} \cdots (w - u_n)^{k_n}\, dw + D, \quad (25.32)$$

where u_i represents a specified point on the real w-axis (a turning point) and k_i is related to the interior angle α_i of the polygon by

$$k_i = \frac{\alpha_i}{\pi} - 1. \quad (25.33)$$

$-k_i\pi$ is the turning angle; if $k_i = 0$, there is no turning. (See Fig. 25.3.)

The lower limit of integration determines what point maps into the origin in the a-plane; it can be replaced by the additive constant D; both serve to translate the resulting polygon. The factor C serves to magnify (or contract) the resulting polygon and to rotate it if $Re\{C\} \neq 0$.

Note that infinity in the w-plane maps into a point on the periphery of the polygon. The turning angle for this point will be determined by the other turning angles, so that they all add up to 2π. If the other turning angles themselves add to 2π, then infinity will map into a point on a side of the polygon; otherwise it becomes a corner of the polygon. Note also that

$$(w - a)^{1/2}(w - b)^{1/2} \neq [(w - a)(w - b)]^{1/2} \quad (25.34)$$

because of the method of selecting the principal values of the square root. Also, negative turning angles are permitted.

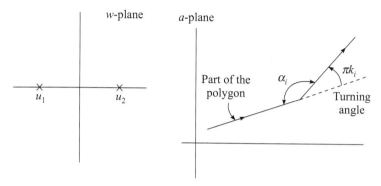

Figure 25.3 Constructing a polygon (in the a-plane) with the Schwarz–Christoffel transformation.

Let us see how this works, with a specific transformation, related to our solving the original problem of steady heat conduction in a tall rectangle.

$$a = \frac{-1}{\sqrt{2}} \int_{-1}^{w} \frac{dw}{(w-1)(w+1)^{1/2}}. \tag{25.35}$$

We can look up this particular integral, but first we want to contemplate what is happening from the point of view of the general Schwarz–Christoffel transformation.

−1 on the real w-axis maps into the origin in the a-plane, because this is the lower limit of integration and there is no additive constant (like D). To determine the direction that the line (for the real w-axis) takes, look at the factors under the integral. Between A and B, $w-1$ is negative, $w+1$ is positive, and $(w+1)^{1/2}$ is positive. Together with the factor $-1/\sqrt{2}$ in front,

$$\frac{-1}{\sqrt{2}} \frac{1}{w-1} \frac{1}{(w+1)^{1/2}}$$

is a positive factor. Thus da/dw is positive, or da is positive if dw is positive. The mapping of the AB line in the w-plane extends horizontally from the origin in the a-plane. (See Fig. 25.4.)

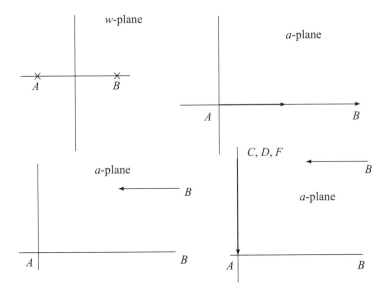

Figure 25.4 Scoping out the mapping in the *a*-plane. First determine the direction and placement of the *AB* line in the *a*-plane. Then determine the direction after the turn at *B*. Next look at the *DA* line to determine how far up (in the a_i direction) to place the *BC* line.

How far do we go in this direction before we change direction? The factor (direction) doesn't change until we reach B (or u_2) if we proceed along the *u*-axis. $1/(w - 1)$ goes to infinity as *u* approaches 1, and it does so in a nonintegrable manner. Thus, the point B gets pushed off to infinity. The transformation has a singularity there.

When we cross $u = u_2 = 1$, the direction of da/dw changes. Only one factor in the integrand of *a* changes when we cross the point $u = u_i$. Therefore the exponent k_i on this factor determines the change in direction. Let us look at our situation when we cross u_2. $w-1$ changes sign to positive. $w +1$ and $(w + 1)^{1/2}$ remain positive. The sign of *da* is now negative if *dw* is positive. We are coming back from infinity. The turning angle is π: $k_2 = -1$; the turning angle is $-k_i\pi = \pi$. The interior angle is $\alpha_2 = 1 + \pi k_2 = 0$. (This is also shown in Fig. 25.4.)

But we do not know exactly where we are, that is, how large is Im{*a*} or a_i.

Let us analyze the situation to the left of $u_1 = -1$. Here $w - 1$ is negative, $w + 1$ is negative, $(w + 1)^{1/2}$ is proportional to *i*, and $1/(w + 1)^{1/2}$ is proportional to $-i$. Thus, da/dw is proportional to

–i. If dw is positive, da is proportional to –i. The direction of DA is downward, as shown in the bottom right sketch in Fig. 25.4.)

How far does this go in the a_i direction? We have to integrate from minus infinity:

$$a_D = \frac{-1}{\sqrt{2}} \int_{-1}^{-\infty} \frac{dw}{(w-1)(w+1)^{1/2}} = \frac{-1}{\sqrt{2}} \int_{-1}^{-\infty} \frac{du}{(u-1)(u+1)^{1/2}}. \qquad (25.36)$$

Make the substitution of U for $-u$, and get

$$a_D = \frac{-1}{\sqrt{2}} \int_{1}^{\infty} \frac{dU}{(U+1)i(U-1)^{1/2}} = \frac{i}{\sqrt{2}} \int_{1}^{\infty} \frac{dU}{(U+1)(U-1)^{1/2}} = i\frac{\pi}{2}. \qquad (25.37)$$

The mapping from the w-plane to the a-plane is now clear, and is shown in Fig. 25.1. The boundary conditions are also shown on the figure. Since T satisfies Laplace's equation in each coordinate system, it becomes apparent that the solution for T is

$$T = T_0\left(1 - \frac{2}{\pi}a_i\right). \qquad (25.38)$$

We still need to see what this result means in the original coordinate system, and how the flux varies along the bottom and right sides of the original region of interest.

The analytic form for the transformation, obtained by looking up the integral in a table, is

$$a = -\frac{1}{2}\ln\frac{\sqrt{2}-\sqrt{1+w}}{\sqrt{2}+\sqrt{1+w}}. \qquad (25.39)$$

Since $w = -\cos(\pi z/L)$ and $1 + w = 2\sin^2(\pi z/2L)$, this can be related all the way back to the original variable z

$$a = -\frac{1}{2}\ln\left(\frac{1-\sin(\pi z/2L)}{1+\sin(\pi z/2L)}\right). \qquad (25.40)$$

You can even write down the temperature profile in the original coordinate system (see Eq. 15.34).

25.6   Translating the Flux Densities

In Chapter 15 are given the flux densities on the right and lower boundaries, as obtained directly from the temperature distribution in Eq. 15.34. Let us see how to obtain these results in a more direct way from the transformations. This is especially useful if we do not need the values of $T(x, y)$ itself.
Let

$$\phi = S + iT = \frac{2T_0}{\pi}\left(i\frac{\pi}{2} - a\right). \tag{25.41}$$

This is constructed so that T is the imaginary part of ϕ but so that ϕ is a simple complex function of a. We can see by inspection how to do this so that ϕ depends only on a but not separately on the real and imaginary parts of a. You can also do this by constructing S, the real part of ϕ, so that S and T satisfy the Cauchy–Riemann conditions (see Problem 25.5).

Remember that $d\phi/dz$ is related to the derivatives of T in the z-plane:

$$\frac{d\phi}{dz} = \frac{\partial T}{\partial y} + i\frac{\partial T}{\partial x} \tag{25.42}$$

(see Eq. 25.13). This is helpful in constructing the flux components.

It is valid to use the chain rule of differentiation for complex functions. If we want the derivative $d\phi/dz$, because we want derivatives of T in the z-plane, then we write

$$\frac{d\phi}{dz} = \frac{\partial\phi}{\partial a}\frac{da}{dw}\frac{dw}{dz}. \tag{25.43}$$

We know all the derivatives on the right side of the equation, $d\phi/da$ because we constructed it to give the solution of Laplace's equation in the a-plane and the other two because they constitute derivatives used to construct the two transformations. Thus,

$$\frac{d\phi}{da} = \frac{2T_0}{\pi}, \quad \frac{da}{dw} = \frac{-1/\sqrt{2}}{(w-1)(w+1)^{1/2}}, \quad \text{and} \quad \frac{dw}{dz} = \frac{\pi}{L}\sin\left(\frac{\pi z}{L}\right).$$

$$\tag{25.44}$$

Suppose we want derivatives on AB (that is, on $y = 0$, $0 < x < L$) in the z-plane. This corresponds to $z = x$ in the z-plane, $\upsilon = 0$, and $w = u$ for $-1 < u < 1$ in the w-plane, giving

$$\frac{dw}{dz} = \frac{\pi}{L}\sin\left(\frac{\pi x}{L}\right) \quad \text{and} \quad \frac{da}{dw} = \frac{-1}{\sqrt{2}}\frac{1}{(u-1)(u+1)^{1/2}} \quad (25.45)$$

so that

$$\frac{d\phi}{dz} = -\frac{2T_0}{\pi}\left(\frac{-1}{\sqrt{2}}\right)\frac{1}{(u-1)(u+1)^{1/2}}\frac{\pi}{L}\sin\left(\frac{\pi x}{L}\right). \quad (25.46)$$

A little trigonometry shows that

$$u = -\cos\left(\frac{\pi x}{L}\right) \quad (25.47)$$

and

$$u - 1 = -2\cos^2\left(\frac{\pi x}{2L}\right), \quad (u+1)^{1/2} = \sqrt{2}\sin\left(\frac{\pi x}{2L}\right),$$

$$\text{and} \quad \sin\left(\frac{\pi x}{L}\right) = 2\sin\left(\frac{\pi x}{2L}\right)\cos\left(\frac{\pi x}{2L}\right). \quad (25.48)$$

Putting it all together, we have

$$\frac{d\phi}{dz} = -\frac{T_0}{L}\frac{1}{\cos(\pi x/2L)}, \quad (25.49)$$

a real quantity. From Eq. 25.42, we see that this is the desired quantity and therefore

$$\left.\frac{dT}{dy}\right|_{y=0} = -\frac{T_0}{L}\frac{1}{\cos(\pi x/2L)}, \quad (25.50)$$

and $q_y(x, 0)$ follows (see Eq. 15.35).

25.7 Transformations for a Photoelectrochemical Cell

The Schwarz–Christoffel transformation is a design problem. How should one obtain the parameters (u_i, k_i, C, and D) to get the desired polygon? The shape of the polygon makes the choice of k_i clear because the turning angles determine k_i uniquely. The choice of u_i is more difficult. One can choose two of these, u_1 and u_2, more or less arbitrarily, because they determine the origin and the length of one side. These can be adjusted later, if desired, by adjusting C and D, because C and D expand, rotate, and translate the polygon. For a complex figure, choosing u_3, u_4, ... can be a trial-and-error procedure, possibly using a computer.

As an example, the geometry in Fig. 25.5 arises in a photoelectrochemical cell. If we ignore the concentration gradients and electrode kinetics, we may be content to estimate the resistance of the cell [1]. Then AB and EF are equipotential electrodes, and the other sides are insulators.

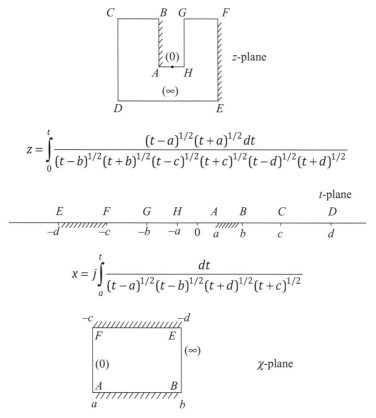

$$z = \int_0^t \frac{(t-a)^{1/2}(t+a)^{1/2}\,dt}{(t-b)^{1/2}(t+b)^{1/2}(t-c)^{1/2}(t+c)^{1/2}(t-d)^{1/2}(t+d)^{1/2}}$$

$$x = j\int_a^t \frac{dt}{(t-a)^{1/2}(t-b)^{1/2}(t+d)^{1/2}(t+c)^{1/2}}$$

Figure 25.5 The z-plane shows a symmetry section of a photoelectrochemical cell. AB and EF are electrodes. The other sides are insulators. The t-plane is after mapping the cell into the upper half plane. The χ-plane is after wrapping into a rectangle with a known solution.

To get a transformation from the z-plane to the t-plane, one needs to obtain u_i. The problem was made a little simpler by requiring that $BC = FG$ and $AB = GH$. This left b, c, and d to be determined.

Going from the t-plane to the χ-plane is straightforward, because the values of the u_i and k_i are now determined. (We use a different set of parameters for the second transformation.) To calculate the resistance requires determining the current distribution on AB (or EF) and integrating to determine the total current.

Problems

25.1 Write out the real and imaginary parts (u and v) of $w = f(z)$ from the definition in Eq. 25.6. You should get

$$u = -\cos\left(\frac{\pi x}{L}\right)\cosh\left(\frac{\pi y}{L}\right) \quad \text{and} \quad v = \sin\left(\frac{\pi x}{L}\right)\sinh\left(\frac{\pi y}{L}\right).$$

(25.51)

25.2 Recall that in Chapter 14, we determined the general solution of Laplace's equation.

We had never seen any real application of that general solution applied to an elliptic equation. Establish a relationship between the function $f(z)$ and its real and imaginary parts, u and v, and the general solution stated in Chapter 14.

25.3 We are told that we can obtain the conjugate harmonic function v to a given harmonic function u by integrating the Cauchy–Riemann equations. Show that, except for an arbitrary additive constant, this procedure yields a unique value for v, that is, the result is independent of the path of integration. (Compare with the discussion of Stokes's law and the integration of a gradient around a closed contour.)

25.4 Show that the conformal mapping $w = L/z$ transforms straight lines of constant y into circles by showing that

$$u^2 + v^2 \approx \frac{1}{4}\exp(2\pi y/L).$$

(25.52)

25.5 Take T to be defined by Eq. 25.38. Construct S so that S and T are the real and imaginary parts, respectively, of the complex variable ϕ. Do this by integrating the Cauchy–Riemann conditions in an appropriate manner.

25.6 Use the development of the derivative $d\phi/dz$ to obtain the heat flux $q_x(L, y)$ on the right boundary of the rectangle. See Eq. 15.36.

25.7 Show that lines of constant values of a harmonic function are perpendicular to lines of constant values of the conjugate harmonic function, thus substantiating the statement that equipotentials and current lines are specified by two conjugate functions.

Additional Notes

If $f(w)$ is analytic in a region, $\oint f(w)dw = 0$. This says that we can follow any path in integrating Eq. 25.32 or 25.35 to obtain a.

Reference

1. Mark E. Orazem and John Newman, "Primary current distribution and resistance of a slotted-electrode cell," *Journal of the Electrochemical Society*, **131**, 2857–2861 (1984).

Chapter 26

Calculus of Variations

We were always taught that a straight line is the shortest distance betweent two points (see Fig. 26.1). But can we prove that this is so?

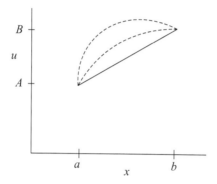

Figure 26.1   Shortest line between two points.

Let us first set up the problem in mathematical terms. The length of a line, curved or not, can be expressed as

$$L = \int_a^b \sqrt{1+(u')^2} \; dx, \qquad (26.1)$$

where $u(x)$ is the y-position of the line, as a function of x, and $u' = du/dx$. This expression comes from the differential length ds:

The Newman Lectures on Mathematics
John Newman and Vincent Battaglia
Copyright © 2018 Pan Stanford Publishing Pte. Ltd.
ISBN 978-981-4774-25-3 (Hardcover), 978-1-315-10885-8 (eBook)
www.panstanford.com

$$ds^2 = dx^2 + dy^2 \quad \text{or} \quad ds = \left[1 + \left(\frac{dy}{dx}\right)^2\right]^{1/2} dx. \qquad (26.2)$$

a and b are the x-values at the left and right ends of the line, and the two points (between which the line is to be drawn) require that the end points be specified as, for example,

$$u(a) = A \text{ and } u(b) = B. \qquad (26.3)$$

The problem of finding the shortest distance between the two points a, A and b, B therefore is to find the minimum value of the above integral L for all possible curves $u(x)$ which pass through the two end points. This may seem like a formidable problem, but at least we now have it stated in mathematical terms. The way of approaching such problems has been formulated and is called the *calculus of variations*. The methodology leads to the Euler–Lagrange formula, as developed next.

26.1 The Euler–Lagrange Formula

Let

$$u_0 = u + \varepsilon\eta(x), \qquad (26.4)$$

where $u(x)$ is the desired answer, as yet unknown, ε is a parameter for a family of functions defined by $\eta(x)$. That is, η is a specific, but arbitrary, twice differentiable function which is (in the present example) zero at the two end points, $x = a$ and $x = b$, and defines the family of functions.

Let

$$I(\varepsilon) = \int_a^b F(x, u_0, u_0') \, dx, \qquad (26.5)$$

where F is a function of x and also a function of two other functions, u_0 and u_0'. I believe that such a thing gets the name *functionale*, but you should check this.

Our objective now is to obtain a minimum (or a maximum) of I by selecting the value of ε which makes the derivative with respect to ε equal to zero. In addition, if we have obtained the correct function for $u(x)$, then ε will also be zero at the extremum. We learned back in Chapter 1 how to differentiate an integral involving a function. Following this carefully, we obtain

$$\frac{dl}{d\varepsilon} = \int_a^b \left[\frac{\partial F}{\partial u} \frac{du_0}{d\varepsilon} + \frac{\partial F}{\partial u'} \frac{du'_0}{d\varepsilon} \right] dx. \qquad (26.6)$$

Following the definition 26.4 gives

$$\frac{dl}{d\varepsilon} = \int_a^b \left[\frac{\partial F}{\partial u} \eta + \frac{\partial F}{\partial u'} \frac{d\eta}{dx} \right] dx. \qquad (26.7)$$

As observed earlier, we want this derivative to be zero. Let us integrate the second term by parts, yielding

$$\int_a^b \frac{\partial F}{\partial u'} \frac{d\eta}{dx} dx = \frac{\partial F}{\partial u'} \eta \Big|_a^b - \int_a^b \frac{d}{dx} \left(\frac{\partial F}{\partial u'} \right) \eta dx. \qquad (26.8)$$

In the simple example being considered here, $\eta = 0$ at both $x = a$ and $x = b$. Thus, the two terms in Eq. 26.7 together give

$$\int_a^b \left[\frac{\partial F}{\partial u} - \frac{d}{dx} \left(\frac{\partial F}{\partial u'} \right) \right] \eta dx = 0 \qquad (26.9)$$

Since this must be true for any arbitrary function η, the term in brackets must be zero over the whole range from a to b:

$$\frac{\partial F}{\partial u} - \frac{d}{dx} \left(\frac{\partial F}{\partial u'} \right) = 0. \qquad (26.10)$$

This is called the Euler–Lagrange equation. With a little work, it also provides a differential equation for determing u.

26.2 Application to the Shortest Distance

We still need to figure out how to apply the Euler–Lagrange equation. For the problem of finding the shortest distance between two points,

$$F = \sqrt{1 + (u')^2}, \qquad (26.11)$$

and thus does not depend explicitly on either u or x, and we have

$$\frac{\partial F}{\partial u} = 0 \quad \text{and} \quad \frac{\partial F}{\partial u'} = \frac{1}{2} \frac{2u'}{\sqrt{1 + (u')^2}}. \qquad (26.12)$$

With a little more effort, we can determine that

$$\frac{d}{dx} \frac{\partial F}{\partial u'} = \frac{u''}{\sqrt{1 + (u')^2}} - \frac{u'}{2} \frac{2u'u''}{[1 + (u')]^{3/2}}. \qquad (26.13)$$

We can factor out u'' and put the rest over a common denominator

$$\frac{d}{dx}\frac{\partial F}{\partial u'} = u''\frac{1+(u')^2-(u')^2}{[1+(u')^2]^{3/2}} = \frac{u''}{[1+(u')^2]^{3/2}}. \qquad (26.14)$$

Thus, the Euler–Lagrange equation reduces, in this case, to

$$u'' = 0, \qquad (26.15)$$

with the solution

$$u = A + (B-A)\frac{x-a}{b-a}, \qquad (26.16)$$

a straight line between the two points.

As noted earlier, the Euler–Lagrange equation provides a differential equation for determining u. Since such differential equations are more familiar to us than determining a function by minimizing an integral, we feel more comfortable by having reduced this to an ordinary differential equation, even if we have to solve the differential equation numerically.

26.3 Optimization with a Constraint

The simplest type of extremum problem would include the problem just treated, with fixed end points, but also with a more complicated function $F(x,u,u')$. The second level of complexity involves having a constraint. An example would be the catenary, where the question is to find the shape of a limp cord draped between two points, but with a length greater than the shortest distance between the two points (see Fig. 26.2). This can be expressed mathematically as follows. Minimize the potential energy

$$E = \int_a^b \sqrt{1+(u')^2}\,mgu\,dx \qquad (26.17)$$

with $u(a) = A$ and $u(b) = B$ and subject to the constraint that the curve is of fixed length

$$L = \int_a^b \sqrt{1+(u')^2}\,dx. \qquad (26.18)$$

Here m is the mass per unit length of the cord, and g is the gravitational acceleration. Both of these are constants.

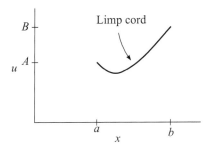

Figure 26.2 Catenary, the shape of a limp cord suspended at the ends.

We proceed now by minimizing

$$E - \lambda L = \int_a^b \sqrt{1 + (u')^2}\, (mgu - \lambda)\, dx, \qquad (26.19)$$

where λ is an undetermined constant. This means that we have multiplied the constraint by λ and subtracted it from the potential energy. We then have minimized the resulting integral.

With some effort, the readers should be able to derive the following differential equation using the Euler–Lagrange equation:

$$mg[1 + (u')^2] = (mgu - \lambda)u'' \qquad (26.20)$$

with the boundary conditions given above.

The procedure is to get a solution for u as a function of x with λ as a parameter: $u = u(x, \lambda)$. Substitute this back into the constraint to solve for λ

$$L = \int_a^b \sqrt{1 + [\partial u(x, \lambda)/\partial x]^2}\, dx. \qquad (26.21)$$

You can appreciate that you might need a numerical solution of the differential equation and a numerical search for the correct value of λ. In some classical problems, like the catenary, the old timers may have developed analytic solutions.

26.4   Optimization with End Points Not Fixed

When a drop of liquid of given volume is placed on a rigid surface, the end points a and b are not specified; instead the contact angle of the liquid with the solid is given. This is the problem of a sessile drop (see Fig. 26.3). How should we determine both the contact angles and the shape of the droplet?

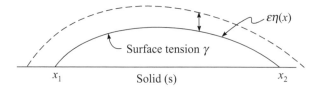

Figure 26.3 Sessile drop.

The problem is formulated as one of minimizing the energy E, composed of surface energies and possibly other energies such as gravitational and thin-film forces. The problem with gravitational forces alone, but not thin-film forces, is a classical one. We formulate here the one with thin-film forces (see Refs. 1–3).

We want to minimize E, where

$$E = \gamma_{\mathrm{SL}} \mathcal{A}_{\mathrm{SL}} + \gamma_{\mathrm{SO}} \mathcal{A}_{\mathrm{SO}} + W \int_{x_1}^{x_2} [\gamma ds + P(h)\, dx]. \qquad (26.22)$$

Here γ_{SL}, γ_{SO}, and γ are the surface energies per unit area (or the surface tensions) of the interfaces between the solid and the liquid, the solid and the vapor (or vacuum), and the liquid and the vapor, respectively. $\mathcal{A}_{\mathrm{SL}}$ and $\mathcal{A}_{\mathrm{SO}}$ are the areas of the interfaces between the solid and the liquid or vapor, respectively, and the integral over ds represents the area of the liquid–vapor interface. The thin-film energy is represented by $P(h)$ and signifies an interaction between the solid–liquid interface and the liquid–vapor interface, acting through the intermediate fluid. This energy should decrease toward zero as the separation distance h increases. h is thus the position of the latter interface; knowing its value as a function of x specifies the shape of the droplet. W is perpendicular to the plane of x and h.

Let us use the variable u instead of h, in keeping with the notation used in the earlier examples. Then

$$ds = \sqrt{1 + (u')^2}\, dx. \qquad (26.23)$$

Also,

$$d\mathcal{A}_{\mathrm{SL}} + d\mathcal{A}_{\mathrm{SO}} = 0 \qquad (26.24)$$

since the area of the solid is fixed, and

$$dA_{SL} = Wd(x_2 - x_1). \tag{26.25}$$

Assume symmetry, and treat only half of the system.

To be more realistic, we should use an axisymmetric droplet instead of a cylindrical one [1–3]. Theoreticians prefer cylindrical droplets, despite their lack of physical significance, because at a later stage one can use reduction in order (see Chapter 5).

Thus, here we want to minimize

$$E = W \int_{x_1}^{x_2} \left[\gamma \sqrt{1+(u')^2} + P(u) + \gamma_{SL} - \gamma_{SO} \right] dx + \text{constant} \tag{26.26}$$

subject to the volume constraint that

$$Q = W \int_{x_1}^{x_2} u \, dx \text{ is a constant.} \tag{26.27}$$

Thus, minimize $E - \lambda Q$.

Now

$$u_o = u + \varepsilon \eta(x), \tag{26.28}$$

but η is not necessarily zero at $x = x_1$ and $x = x_2$. This means that we have to go back to the derivation of the Euler–Lagrange equation and focus on the integration by parts. The equation becomes (see Section Additional Notes)

$$\frac{1}{2W} \frac{\delta(E - \lambda Q)}{\delta \varepsilon} = 0 = \int_0^{x_2} \left[-\gamma \frac{u''}{\left[1+(u')^2\right]^{3/2}} + \frac{dP}{du} - \lambda \right] \eta \, dx$$

$$+\eta(x_2) \left[\gamma \frac{u'}{[1+(u')^2]^{1/2}} - \frac{1}{u'} (\gamma[1+(u')^2]^{1/2} + P(0) + \gamma_{SL} - \gamma_{SO}) \right]_{x_2}. \tag{26.29}$$

Both coefficients of η must be zero, leading first to an augmented Young–Laplace equation

$$\frac{u''}{[1+(u')^2]^{3/2}} = \frac{1}{\gamma} \left(\frac{dP}{du} - \lambda \right). \tag{26.30}$$

(Without dP/du, this gives a circular arc; see Problem 26.1.) The Young–Laplace equation is a differential equation describing the shape of a bubble or drop. The word *augmented* is added to cover the inclusion of the term with dP/du.

The boundary condition arises from setting the coefficient of the second term in Eq. 26.29 to zero.

$$\frac{\gamma}{[1+(u')^2]^{1/2}}+P(0)+\gamma_{SL}-\gamma_{SO}=0 \quad \text{at} \quad x=x_2. \quad (26.31)$$

This can be expressed in terms of the *contact angle* θ as

$$\gamma|\cos\theta|=\gamma_{SO}-\gamma_{SL}-P(0). \quad (26.32)$$

This is called the augmented Young equation, the augmentation coming from the presence of the term $P(0)$. For very small droplets, the thin-film forces can lead to a change of the shape of the droplet.

26.5   Other Examples

The above examples emphasize phase 1 (set up the problem) and give less attention to phases 2 (solve it) and 3 (contemplate the result).

Examples of problems in the calculus of variations abound in science and technology and in folklore. Try your hand at setting up the problem for a few of the following topics:

(A) Thermodynamic equilibrium implies maximization of the entropy or minimization of the Helmholtz or Gibbs energy, depending on the constraint. Consider how you would show that the temperatures of two thermal reservoirs are the same if they can exchange thermal energy and the entropy is to be maximized.

(B) Find the shape of a liquid on the surface of a wire, where the liquid tends to bead up on the wire. A fluid of a given mass per unit length coats a wire. What shape will it take to minimize the surface area if it is constrained to have a given amplitude and a given wavelength? See Fig. 26.4.

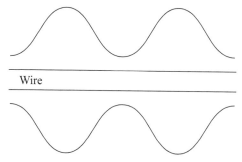

Figure 26.4   Liquid beads on a wire.

(C) Roller coaster design. What should be the shape of a frictionless slide in order for an object to fall from one fixed point to another in the minimum time?

(D) In the days of Queen Dido of Carthage, someone was offered as much land, adjacent to an irregular coastline, as he could plow around in one day. How should he plan his day so as to acquire the most land and still finish before sunset?

(E) How to go from an Eulerian formulation of mechanics to a Lagrangian formulation, resulting in the conventional form of Newton's second law of motion:

$$F = m\frac{dv}{dt}. \tag{26.33}$$

Sometimes one wonders whether these cute variational formulations express the underlying physical law correctly. The methodology can be used to show the equivalence of the two formulations.

(F) Onsager's principle of minimum dissipation of energy. This has been called into question recently, but it is commonly used in the derivation of the Onsager reciprocal relations.

(G) What should be the altitude profile for an airplane flying from San Francisco to New York in order to minimize the fuel consumption?

Problems

26.1 (a) Show that the cylindrical droplet has the shape of a circular arc if $dP/du = 0$ by integrating Eq. 26.30. This applies to axisymmetric as well as cylindrical drops. Consider what happens with different volumes.

(b) Still with $P(u) = 0$, map out $u(x, \lambda)$ as a function of λ. Calculate the volume of the droplet Q/W as a function of λ and thus come to understand better the meaning of λ.

26.2 Carry out the development of drop shape for axisymmetric droplets.

26.3 Show how to include gravitational effect in the formulation of the drop-shape problem.

26.4 Derive the catenary differential equation from a force balance. The force along the cord must be in the direction of the cord (thus the meaning of the word *limp*). (See Fig. 26.5.)

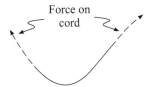

Figure 26.5   For the profile of a limp cord, the force exerted by adjacent parts must be tangent to the direction of the cord.

26.5 (a) Develop the solution for the catenary curve.

(b) Nondimensionalize the problem and seek a critical parameter for the shape of a catenary.

26.6 If

$$I = \int_a^b F(x; u_1, u_2 \dots, u_n; u_1', u_2' \dots u_n') \, dx \qquad (26.34)$$

for n independent unknown functions $u_k(x)$, show that a similar Euler–Lagrange equation is obtained for each u_k, that is,

$$\frac{\partial F}{\partial u_k} - \frac{d}{dx}\left(\frac{\partial F}{\partial u_k'}\right) = 0, \quad \text{for} \quad k = 1, 2, \dots, n. \qquad (26.35)$$

Additional Notes

Derivation for Unfixed Endpoints

Starting with Eq. 26.29, we need to include more detail.

$$\frac{1}{2W}\frac{\delta(E-\lambda Q)}{\delta\varepsilon}=0=\frac{\delta}{\delta\varepsilon}\int_0^{x_2}\left[\gamma\sqrt{1+(u')^2}+P(u)+\gamma_{SL}-\gamma_{SO}-\lambda u\right]dx$$

$$=\frac{\delta}{\delta\varepsilon}\int_0^{x_2}F(x,u_0,u_0')dx=\int_0^{x_2}\left[\frac{\partial F}{\partial u}\frac{\partial u_0}{\partial\varepsilon}+\frac{\partial F}{\partial u'}\frac{\partial u_0'}{\partial\varepsilon}\right]dx+\frac{\delta x_2}{\partial\varepsilon}F\bigg|_{x_2}.$$

$$(26.36)$$

With Eq. 26.4, the terms in the integral are treated as before. From Fig. 26.6 one can infer that

$$u_0'(x_2)\approx-\frac{\delta\varepsilon\eta(x_2)}{\delta x_2}\quad\text{or}\quad\frac{\delta x_2}{\delta\varepsilon}=-\frac{\eta(x_2)}{u'(x_2)}.\qquad(26.37)$$

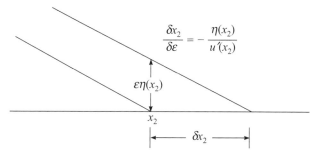

Figure 26.6 Near the edge of the droplet, the slope of the profile determines the variation of x_2 with ε.

The expression then becomes

$$0=\int_0^{x_2}\left[\frac{\partial F}{\partial u}\eta+\frac{\partial F}{\partial u'}\frac{d\eta}{dx}\right]dx-\frac{\eta}{u'}F\bigg|_{x_2}.\qquad(26.38)$$

As before, the second term in the integral is to be integrated by parts, but this time we cannot set η equal to zero at the end points. We obtain

$$0=\int_0^{x_2}\left[\frac{\partial F}{\partial u}\eta dx+\eta\frac{\partial F}{\partial u'}\right]_0^{x_2}-\int_0^{x_2}\eta\frac{d}{dx}\left(\frac{\partial F}{\partial u'}\right)dx-\frac{\eta}{u'}F\bigg|_{x_2}$$

$$=\int_0^{x_2}\left[\frac{\partial F}{\partial u}-\frac{d}{dx}\left(\frac{\partial F}{\partial u'}\right)\right]\eta dx+\eta\left[\frac{\partial F}{\partial u'}-\frac{1}{u'}F\right]_{x_2}.$$

$$(26.39)$$

The consequence of η being arbitrary is that the bracketed term in the integrand must be zero (as before)

$$\frac{\partial F}{\partial u} - \frac{d}{dx}\left(\frac{\partial F}{\partial u'}\right) = 0, \tag{26.40}$$

and the boundary condition becomes

$$\frac{\partial F}{\partial u'} - \frac{1}{u'}F = 0 \quad \text{at} \quad x = x_2. \tag{26.41}$$

In this problem,

$$F = \gamma\sqrt{1+(u')^2} + P(u) + \gamma_{SL} - \gamma_{S0} - \lambda u.$$

Hence,

$$\frac{\partial F}{\partial u} = \frac{dP}{du} - \lambda, \tag{26.42}$$

and

$$\frac{\partial F}{\partial u'} = \frac{\gamma}{2}\frac{2u'}{\sqrt{1+(u')^2}}. \tag{26.43}$$

(We notice that $\partial F/\partial u' = 0$ at the point of symmetry at $x = 0$.) Further differentiation gives

$$\frac{d}{dx}\left(\frac{\partial F}{\partial u'}\right) = \gamma\frac{u''}{\sqrt{1+(u')^2}} - \frac{\gamma u'}{2}\frac{2u'u''}{[1+(u')^2]^{3/2}}$$

$$= \gamma u''\frac{1+(u')^2-(u')^2}{[1+(u')^2]^{3/2}} = \frac{\gamma u''}{[1+(u')^2]^{3/2}}. \tag{26.44}$$

Hence, the Euler–Lagrange equation becomes

$$\frac{u''}{[1+(u')^2]^{3/2}} = \frac{1}{\gamma}\left(\frac{dP}{du} - \lambda\right), \tag{26.45}$$

which is Eq. 26.30, the augmented Young–Laplace equation. The boundary condition becomes

$$\frac{\gamma u'}{\sqrt{1+(u')^2}} = \frac{1}{u'}\left[\gamma\sqrt{1+(u')^2} + P(u) + \gamma_{SL} - \gamma_{S0} - \lambda u\right] \quad \text{at} \quad x = x_2. \tag{26.46}$$

But $u = 0$ at $x = x_2$, and the other terms can be combined to give

$$\frac{\gamma}{\sqrt{1+(u')^2}}\left[(u')^2 - 1 - (u')^2\right] = P(0) + \gamma_{SL} - \gamma_{S0} \quad \text{at} \quad x = x_2 \tag{26.47}$$

or

$$\frac{\gamma}{\sqrt{1+(u')^2}} + P(0) + \gamma_{SL} - \gamma_{S0} = 0 \quad \text{at} \quad x = x_2, \qquad (26.48)$$

the augmented Young equation 26.31 or 26.32.

Catenary

It may be helpful to have additional detail on how one would continue a problem to get a result. Work with the catenary, where the differential equation, after minimization of the potential energy subject to the constraint of fixed cord length, is

$$1 + (u')^2 = (u - \lambda/mg)u'' \qquad (26.49)$$

with the boundary conditions

$$u(a) = A \text{ and } u(b) = B. \qquad (26.50)$$

Since x does not appear explicitly, reduction in order can be used to solve the problem.

What are the critical parameters here? It seems that all catenaries have pretty much the same shape. They can be scaled differently. Already we can see that mg can be combined with the unknown parameter λ. Since u and x have the same dimension (length), u' is dimensionless, and so is uu''.

On physical grounds, we can see that only the differences $b - a$ and $B - A$ are significant, since the resulting figure can be translated. Hence, it could be economical to use the redefined variables

$$U = u - A \quad \text{and} \quad X = x - a. \qquad (26.51)$$

From another point of view, the bottom of the draped section is the most important. For a given catenary, the suspension points a, A and b, B can be placed anywhere along the curve. $-\lambda/mg$ has the dimension of a length and can be regarded to scale the curvature u'' at the bottom of the curve. That is, at the bottom of the curve, $u' = 0$ and $u'' = 1/(u - \lambda/mg)$. If these values (at the bottom of the curve) were used to scale lengths (in both the u- and x-direction), then all catenaries could be superposed on one figure. Depending on the placement of the suspension points, the cord between these points may or may not have a minimum.

Recast in terms of X and U, the problem is

$$1 + \left(\frac{dU}{dX}\right)^2 = (U - K_1)\frac{d^2U}{dX^2}, \tag{26.52}$$

with the boundary conditions

$$U = 0 \quad \text{at} \quad X = 0 \quad \text{and} \quad U = B - A \quad \text{at} \quad X = b - a, \tag{26.53}$$

and where

$$-K_1 = A - \lambda/mg. \tag{26.54}$$

The readers should verify that the solution is

$$U = K_1 + l\cosh\left(\frac{X - k_2}{l}\right), \tag{26.55}$$

where l and k_2 are two constants. By inspection, you can see that k_2 is the value of X where the curve has a minimum, and $l = 1/u''$ at this minimum.

A way to proceed is to take l to be known. Then K_1 and k_2 can be solved for (perhaps numerically) so that U satisfies the boundary conditions. Finally, the length of the cord, between the suspension points can be determined from the constraint,

$$L = \int_a^b \sqrt{1 + (u')^2}\,dx. \tag{26.56}$$

With the above solution for U, this gives

$$L = l\left[\sinh\left(\frac{b - a - k_2}{l}\right) - \sinh\left(\frac{-k_2}{l}\right)\right]. \tag{26.57}$$

Again, l scales the lengths. If one considered L specified instead of l, then another trial-and-error solution is necessary.

To implement this, express the boundary conditions as

$$0 = K_1 + l\cosh\left(\frac{-k_2}{l}\right) \quad \text{at} \quad X = X_a = 0 \tag{26.58}$$

and

$$B - A = K_1 + l\cosh\left(\frac{b - a - k_2}{l}\right) \quad \text{at} \quad X = X_b = b - a. \tag{26.59}$$

Eliminate K_1 to get a single equation to solve for k_2:

$$B - A = l\left[\cosh\left(\frac{b-a-k_2}{l}\right) - \cosh\left(\frac{k_2}{l}\right)\right]. \qquad (26.60)$$

After determining k_2, obtain K_1 from

$$K_1 = -l\cosh\left(\frac{k_2}{l}\right). \qquad (26.61)$$

Finally, calculate L

$$L = l\left[\sinh\left(\frac{b-a-k_2}{l}\right) + \sinh\left(\frac{k_2}{l}\right)\right]. \qquad (26.62)$$

The table gives certain key results, and the figure shows the catenary curves between the two separation points. For these results, $b - a = 1$, and $B - A = 1$.

From these results, we observe that L approaches $\sqrt{2}$, the shortest distance between the two suspension points, as $l \to \infty$. The cord is then taut. At the same time, the minimum point, k_2, is no longer between the two suspension points for l greater than about 0.7.

Table 26.1 Calculated results for catenaries

l	0.15	0.2	0.5	1	2	5	10
k_2	0.46461	0.41955	0.11403	-0.35246	-1.24808	-3.90098	-8.31079
K_1	-1.66391	-0.82704	-0.51306	-1.06276	-2.40223	-6.60054	-13.6569
L	4.31682	2.61855	1.54308	1.44436	1.42622	1.41539	1.41451

Note that if $A = B$, $k_2 = (b - a)/2$.

Note also that if we plot $[u - u(x_0)]/l$ *versus* $(x - x_0)/l$, where x_0 is the minimum point equal to k_2, all the catenaries fall on a single curve. This might not be so convenient since the suspension points would be in different places for different values of l.

Reduction in order for the catenary

The governing equation is

$$1 + (u')^2 = (u - \lambda/mg)u''. \qquad (26.63)$$

Let $p = u'$, so that

$$u'' = \frac{dp}{dx} = \frac{dp}{du}\frac{du}{dx} = p\frac{dp}{du} = \frac{1}{2}\frac{dp^2}{du}. \qquad (26.64)$$

Substitute into the differential equation

$$1 + p^2 = (u - \lambda / mg) \frac{1}{2} \frac{dp^2}{du} \qquad (26.65)$$

or

$$\frac{2du}{u - \lambda / mg} = \frac{dp^2}{1 + p^2}. \qquad (26.66)$$

Integration gives

$$2\ln(u - \lambda/mg) = \ln(1 + p^2) + 2\ln l, \qquad (26.67)$$

where $2\ln l$ is the integration constant. We now have

$$1 + p^2 = \left(\frac{u - \lambda / mg}{l}\right)^2 \qquad (26.68)$$

or

$$p^2 = \left(\frac{u - \lambda / mg}{l}\right)^2 - 1 \qquad (26.69)$$

or

$$\frac{du}{dx} = p = \pm\left[\left(\frac{u - \lambda / mg}{l}\right)^2 - 1\right]^{1/2}. \qquad (26.70)$$

Use the plus sign to the right of the minimum; use the negative sign to the left. Simplify things by letting

$$\frac{u - \lambda / mg}{l} = y, \qquad (26.71)$$

so that we have

$$l\frac{dy}{dx} = \pm\sqrt{y^2 - 1}, \qquad (26.72)$$

or

$$\frac{dy}{\sqrt{y^2 - 1}} = \pm\frac{dx}{l}. \qquad (26.73)$$

Look up the integral in a table.

$$\pm\frac{x - k_2}{l} = \ln(y + \sqrt{y^2 - 1}) = \cosh^{-1}(y). \qquad (26.74)$$

The solution becomes

$$y = \cosh\left(\frac{x - k_2}{l}\right) = \frac{u - \lambda / mg}{l}.$$ (26.75)

Rewrite this as

$$u - A = K_1 + l\cosh\left(\frac{x - k_2}{l}\right).$$ (26.76)

References

1. E. K. Yeh, John Newman, and C. J. Radke, "Equilibrium configurations of liquid droplets on solid surfaces under the influence of thin-film forces: Part I. Thermodynamics." *Colloids and Surfaces A: Physiochemical and Engineering Aspects,* **156**, 137–144 (1999).

2. E. K. Yeh, John Newman, and C. J. Radke, "Equilibrium configurations of liquid droplets on solid surfaces under the influence of thin-film forces: Part II. Shape calculations." *Colloids and Surfaces A: Physiochemical and Engineering Aspects,* **156**, 525–546 (1999).

3. Divesh Bhatt, John Newman, and C. J. Radke, "Equilibrium force isotherms of a deformable bubble/drop interacting with a solid particle across a thin liquid film." *Langmuir,* **17**, 116–130 (1999).

Further Readings

Francis B. Hildebrand. *Advanced Calculus for Applications.* Englewood Cliffs: Prentice-Hall, 1976. QA 303 H55 1976

Arvind Varma and Massimo Morbidelli. *Mathematical Methods in Chemical Engineering.* New York: Oxford University Press, 1997. TP 149 V36 1997

Erwin Kreyszig. *Advanced Engineering Mathematics.* New York: Wiley. 1972. QA 401 K7 1972 (or 1999 ed. QA 401 K7 1999 in physics library)

C. Ray Wylie and Louis C. Barrett. *Advanced Engineering Mathematics.* New York: McGraw-Hill, 1995. QA 401 W9 1995 (engineering library)

Milton Abramowitz and Irene A. Stegun. *Handbook of Mathematical Functions.* New York: Dover Publications, Inc., 1964. QA47 A4 1964

H. S. Carslaw and J. C. Jaeger. *Conduction of Heat in Solids.* Oxford: Clarendon Press, 1959. QC 321 C28 1959

Ruel V. Churchill. *Operational Mathematics.* New York: McGraw-Hill Book Company, 1958. QA 432 C45 1958

Ruel V. Churchill. *Fourier Series and Boundary Value Problems.* New York: McGraw-Hill Book Company, 1963. QA404C49 1963 (engineering library)

Ruel V. Churchill. *Complex Variables and Applications.* New York: McGraw-Hill Book Company, 1960. QA331 C524 1960

L. I. Sedov. *Similarity and Dimensional Methods in Mechanics.* New York: Academic Press, 1959. QC 39S413

W. F. Ames. *Nonlinear Partial Differential Equations in Engineering.* New York: Academic Press, 1965. QA374A45

Index: Mathematics

Index: Physics

Printed and bound by CPI Group (UK) Ltd, Croydon, CR0 4YY

23/10/2024

01777667-0003